CW00536873

A Sense of Urgency

A Sense of Urgency

HOW THE CLIMATE CRISIS
IS CHANGING RHETORIC

Debra Hawhee

The University of Chicago Press   Chicago and London

The University of Chicago Press, Chicago 60637
The University of Chicago Press, Ltd., London
© 2023 by The University of Chicago
All rights reserved. No part of this book may be used or reproduced in
any manner whatsoever without written permission, except in the case
of brief quotations in critical articles and reviews. For more information,
contact the University of Chicago Press, 1427 E. 60th St., Chicago, IL
60637.
Published 2023
Printed in the United States of America

32 31 30 29 28 27 26 25 24 23   1 2 3 4 5

ISBN-13: 978-0-226-82671-4 (cloth)
ISBN-13: 978-0-226-82678-3 (paper)
ISBN-13: 978-0-226-82677-6 (e-book)
DOI: https://doi.org/10.7208/chicago/9780226826776.001.0001

Library of Congress Cataloging-in-Publication Data

Names: Hawhee, Debra, author.
Title: A sense of urgency : how the climate crisis is changing rhetoric /
Debra Hawhee.
Other titles: How the climate crisis is changing rhetoric
Description: Chicago ; London : The University of Chicago Press, 2023. |
Includes bibliographical references and index.
Identifiers: LCCN 2022051301 | ISBN 9780226826714 (cloth) | ISBN
9780226826783 (paperback) | ISBN 9780226826776 (ebook)
Subjects: LCSH: Climatic changes—Social aspects. | Crises—Social
aspects. | Rhetoric.
Classification: LCC QC902.9 .H39 2023 | DDC 304.2/8—dc23/eng20230118
LC record available at https://lccn.loc.gov/2022051301

♾ This paper meets the requirements of ANSI/NISO Z39.48-1992
(Permanence of Paper).

For Sarah, Seth, and Nora,
who help me feel time

CONTENTS

FIGURES

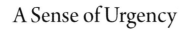

A Sense of Urgency

FRONTISPIECE: Smoky Tennessee sunrise, 2021. Image credit: Adam Gravett.

CHAPTER 1

Introduction

INTENSIFICATIONS

The frontispiece of this book is the work of Adam Gravett, a photographer who frequently treks to the overlook at Kuwahi (also called Clingmans Dome) in the Great Smoky Mountains, the highest point in Tennessee. The huge tree in the foreground casting a stark outline against the typically bluish, misty ridges of the Smokies is a Fraser fir tree (Latin name: *Abies fraseri*), native to the southern Appalachians. It is dead from an infestation of balsam woolly adelgids, small insects that have, over the past few decades, decimated Fraser firs in the area.[1] Gravett captured this image at sunrise on a late July morning in 2021, during a week when residents of Tennessee and North Carolina received air-quality alerts because upper atmospheric air currents had carried dense smoke to the area all the way from Oregon, almost 2,500 miles away, where wildfires had blazed for days.

The visual effects of the smoke are worth dwelling on elementally and figuratively. Elementally, the particle-dense smoke absorbs and bends the shorter blue and violet wavelengths of the sun's light, scattering those wavelengths much more than their longer red and orange counterparts, and so the red and orange wavelengths take on deep, intense hues when sunlight streams through the floating particulate matter. This effect is more pronounced at dawn and dusk, when the sun's oblique angle at the horizon drains the blue from the sky, leaving orange and red wavelengths to catch and radiate through Earth's gaseous atmosphere as they meet the lenses of eyes or cameras. The smoke-dense sunrises and sunsets are at once eerie and astonishing, and those effects gather more significance—they can intensify further—when onlookers learn how far away the smoke originated. It is hard to fathom that smoke particles can travel so far, but of course they can—and farther.

Figuratively, I offer Gravett's photograph in all its disconcerting intensity, as an illustration of how climate change is changing rhetoric, the subject of this book. Like the particles of soot and the jet stream that carries them, rhetoric too drifts and travels, gathers and dissipates; it carries and rides material forces across vast geographic distances, making lived conditions and elsewheres sensible, breathable, harmful—in a word, felt. Much as the smoke from the climate-intensified wildfires deepens the sun's red and orange hues, climate change is also intensifying certain rhetorical concepts, altering perception, and enfolding information within feeling. Knowledge about harmful atmospheric changes produced by humans is spreading, and those changes are often registered on a bodily, felt level—a prickle in the air, a sharp exhale, wide eyes. Feeling meets and mixes with facts and with knowledges to further intensify both.

The idea of intensity will recur throughout this book. At base, *A Sense of Urgency* considers how one intensifier (climate) is altering another kind of intensifier (rhetoric). Climate change is, by most accounts, the grand intensifier. As climate scientists point out, the frequency of wildfires such as the ones in 2021 may remain about the same, but their intensity, duration, and sweep are increasing. So too with hurricanes, floods, and droughts. How such intensification plays out is explained in the first bullet point of the press release for the Intergovernmental Panel on Climate Change (IPCC)'s Landmark Report on the physical science basis of climate change: "Climate change is intensifying the water cycle. This brings more intense rainfall and associated flooding, as well as more intense drought in many regions."[2] Such intensification of floods and drought, and also of fires and storms, has led environmentalists like Hunter Lovins, journalists like Thomas Friedman, and climate scientists like Katharine Hayhoe, also known for her public communication work, to replace the phrase "global warming" with "global weirding."[3] This pivot acknowledges how "warming" loses its usefulness in the context of climate-intensified winter weather resulting from the newly volatile polar vortex, for example.[4] The word "weirding," moreover, names the unexpected, a departure from the ordinary, that which is unusual or unrecognizable, even that which has not happened before. And the word "weird" portends as well. As a noun, it possesses an old association with the Fates.[5] "Global weirding" aptly captures the unpredictable and extreme conditions caused by rising surface temperature as well as the charged strangeness of living through such conditions. The word "weirding" gathers into its colloquial pith a range of intensities that are observable around the world. The frontispiece's eerie image documents but one instance.

Intensification also aptly characterizes the work of rhetoric across the range of materials, knowledge domains, and interventions this book examines. As an art of intensification, rhetoric coils together awareness (knowledges, information) with feeling and casts them across time and between geographic locations. Such a conception of rhetoric unmoors the art from the direct senses of argumentation and persuasion so often attributed to Aristotle and is more akin to that offered by Thomas Farrell, who calls rhetoric "the art, the fine and useful art, of making things matter."[6] An account of rhetoric as an art of intensification burrows more deeply into the feelings involved in making things matter to others. At stake is how "matters of concern" (Bruno Latour's phrase) can be deepened into acts of care.[7] Bringing forward rhetoric as an art of intensification foregrounds quality, attention, and feeling. The Latin word *intendo*, from which the word "intensity" derives, entails a stretching or bending toward something, often of attention (which also shares the same *-tend* root), and it carries valences of magnification and energy.[8] Intensity often characterizes a feeling, not a particular emotion like anger or fear (though it often enlarges these), but strength, force, depth, and density, as for example, in Lisa Corrigan's analysis of the "density of feelings" that gathered in the US civil rights movement.[9] In her account of what she terms "Black feelings," Corrigan observes that "emotional density is a marker of political climate."[10] Corrigan's observation is borne out in the context of the climate crisis as well, as this book will show. Indeed, as an intensifier of already existing disparities and injustices, the changing climate pulls in and worsens other crises and problems—public health crises, racism and racial disparities, social and economic inequalities, settler colonialism—and those convergences further deepen those densities of feeling for particular people.

Geologists, biologists, and climatologists are documenting dramatic changes in sea level, atmospheric composition, habitats, biodiversity, and so on. Scholars from nearly all disciplines (my own included) are documenting injustices wrought by climate change and intransigent governmental policies. With this book, I wish to document changes in rhetoric itself, to gather and consider ways that people are urging each other to imagine their way into, and even out of, an unprecedented set of circumstances. I do so to 1) theorize with more precision the specific rhetorical challenges posed by climate change, and 2) create a blueprint for rhetoric of what Lynn Keller calls "the self-conscious Anthropocene," a term that designates the somewhat recent widespread awareness of the damage our species has done to the planet.[11]

The idea of awareness has long focused efforts to create social change. Awareness-raising, though, is no match for the status quo. The reasons for this mismatch are many and have been chronicled by others:

an abiding (and, for some, profitable) dependency on fossil-based energy; competition for scarce resources; denial; misinformation; finger-pointing; and the apparent geographic distance and creeping pace of some of climate change's effects, what Rob Nixon calls "slow violence."[12] Against these currents, is it possible to ensure that awareness takes hold, effecting the kinds of changes that need to happen? Answers to that question may prove elusive, but examining the novel strategies used by climate rhetors trying to do just that can tell us a good deal about the limits of and possibilities for rhetoric in a rapidly changing climate. Doing so also intensifies, even changes, available rhetorical concepts and practices. Documenting those intensifications—those changes—is the task of this book.

My premise, then, is that rhetoric can change and in fact is already changing in response to the worsening climate crisis. If there is any hope of using the art of intensification (rhetoric) to bend back on the grand intensifier (climate)—to, that is, effect change in the other direction— then it helps to dig into what is happening on the ground rhetorically in response to the rapidly changing climate.[13]

What is happening is this: efforts to imagine futures bring to the fore the rhetorical feature of *magnitude*, winding magnitude with information in order to enlarge the crisis. Such efforts give the climate crisis *presence*, another rhetorical concept this book draws out with a new stress on its sensory and material qualities. Arts and acts of *witnessing*, fortified with the clarifying power of insistence that they gathered over the course of the last century, are expanding to include nonhumans as well as humans, their temporalities unspooling from past, to present, and even to future, and not necessarily in that order. These means of intensification—magnitude, presence, and witnessing—show more precisely how feeling entwines with facts to close the geographic and temporal distances that often work as obstacles to action, particularly for settler descendants in the so-called Global North.

The account of rhetoric this book brings forth is therefore deeply sensuous. Familiar concepts, considered in the context of the climate crisis, are changing; their intensifying work is becoming more palpable. More than attention or awareness, the *quality* of attention and awareness takes on new significance. The account herein is heavy, by which I mean it focuses on how something like rhetorical weight is both borne and conveyed.[14] This book, that is, presents a felt rhetoric. Later in this introduction, I discuss witnessing, magnitude, and presence in more depth and sketch the book's supporting cast of salient concepts, and I also discuss my methods and interpretive approach and preview the rest of the book. But before I do that, a word.

Unprecedented

The word "unprecedented" underwrites the intensifications this book documents. Pressed into the word are assumptions about time, futures, and available actions, all of which concern rhetoric and ought to concern us all. But because of the new, perhaps paradoxically commonplace status of the word "unprecedented," positing anything as genuinely unprecedented entails a kind of audacity that requires explanation. The word "unprecedented" tipped into overuse, by most accounts, when the world found itself in the midst of a global pandemic. As I discuss in chapter 4, those conditions further exposed the extent to which precedent—that which has happened before—has traditionally helped to constitute the future, shaped the ways people see, consider, and imagine futures close and distant.

By most denotative accounts, the unprecedented is the unparalleled; it names conditions not previously known or experienced.[15] By keeping the word in play, I am not arguing that no one has faced an uncertain future before. Such an argument would be flat wrong at best. At its worst, though, such an argument would minimize past atrocities that humans endured and have carried out on each other, atrocities that exceeded all prior atrocities.[16] The challenge here, to borrow phrasing from Elizabeth M. DeLoughrey, is acknowledging novelty while still "being attentive to the historical continuity of dispossession and disaster caused by empire."[17] In other words, Indigenous people who had land taken from them as settler colonists clear-cut their way through the Americas may rightly object that there is not a whole lot new in this moment.[18] Rhetorically, the act of framing something like an event, a period, an era, or a crisis as unprecedented sets it apart, makes it remarkable. And yet far from minimizing particular past atrocities, the most complex instances of climate rhetoric, as this book shows, pull in those atrocities, pull them through time, and hold them together with the unfolding climate crisis.[19]

To say that observable changes in climate are unprecedented is to most climate scientists uncontroversial. The IPCC report published in August 2021 used this exact language in the press release, the technical summary, and the full report.[20] In his account of the current moment, Bruno Latour, building from a 2013 *Le Monde* article by the environmental and climate writer Stéphane Foucart, begins from a premise of the unprecedented:

> Not only do we find ourselves placed at a historic moment without any known precedent ("To find such levels of carbonic gas, we

have to go back to the Pliocene, 2.6 to 5.3 million years ago. The creatures nearest to humans that walked the surface of the Earth at the
time were Australopithecenes!"); not only have we crossed a threshold—a term that is at once legal, scientific, moral, and political; not
only is humanity responsible for this truly revolutionary transformation (this is implied by the well-known association between CO_2
emissions and the industrial way of life); but in addition we have
probably already passed the moment when we could still do something about it.[21]

The use of the word "unprecedented" as a framework for global-political
conditions and as a provable fact are of course two different things,
though both obtain in the context of the climate crisis.[22] To frame climate change—its pace, its magnitude, its measurable conditions—as
"unprecedented" is to place it (and us) on a kind of precipice. The word
carries portents of, at best, a radically uncertain future.[23] Rhetorically,
the word "unprecedented" can put people on edge, because never-
before-known conditions stymie preparation. Protocols do not exist;
they must be created in the moment.

   And yet so many disciplines—medicine, law, and history, to name
a few of the most obvious—rest on a foundation of precedent.[24] Rhetoric, the age-old art of effective communication, is not an exception.
In the Rhetoric, when Aristotle presents deliberative rhetoric—the
future-facing genre in which leaders seek to convince people to take
particular actions—he repeatedly wraps his advice about deliberation
in the soothing tones of precedent: "examples are best in deliberative
speeches; for we divine and judge future things by predicting them
from past ones."[25] Such looping paths back to the past have become
well worn, and they present a quandary in an unprecedented situation,
best formulated as a question: in the absence of precedent, what available means of intensification remain?

   Writing two decades before the arrival of COVID-19, but in the context of turn-of-the-millennium environmental degradation, unchecked
global inequality, and political upheaval, the philosopher and self-
proclaimed "futurist" Jérôme Bindé declared, for example, that "the
past has ceased to be a fund of resources into which the present may
delve in order to prepare for the future."[26] The absence of precedent
is the kind of limitation that scholars, climate activists, scientists, and
artists are trying to overcome with novel, genre-expanding approaches,
mixed and felt temporalities, borrowing rhetorical resources (a familiar
slogan, an image) from another unprecedented situation, and finding
sensuous presence through public art. Finding presence, for particular

groups in particular places, means finding new, imaginative pathways to futures and in places both close and far.

Imagining Futures

The investigative journalist George Monbiot writes toward the end of his 2007 book *Heat* that when it comes to global warming, he himself has "found the likely effects easy to catalogue but almost impossible to imagine."[27] With this observation, Monbiot identifies what environmental scholars from a variety of humanistic disciplines have long considered perhaps the biggest challenge of the climate crisis: the impossibility of fathoming a future where planetary conditions such as atmospheric carbon dioxide, global temperatures, and sea level soar to catastrophic levels. A recent spate of public-facing books written by climate scientists trying to help people grasp why their findings matter, why those findings demand urgent action right now, suggests that the climate crisis is exacerbated by—and perhaps also exacerbates—what the writer Amitav Ghosh diagnoses as a crisis of imagination.[28] For it is through the reaching, ranging work of the imagination that the catalogue of effects comes to matter.

The gulf between cataloguing and imagining, between sheer information and comprehension of information's implications, is created and deepened by the never-beforeness of certain conditions—by, that is, the absence of precedents. Imaginations and futures are stuck together, it would seem; the faculty and the temporality constitute each other. As the anthropologist Cymene Howe, who co-created the ceremony and memorial that I consider in chapter 2, puts it in an interview, "telling time is a matter of imagination."[29] Howe identifies the challenges of "thinking back into the rather unimaginably deep time of rock and magma and tectonic plates," adding that "more difficult still might be the skill of envisioning forward: Casting our sights seven, twelve, seventy-two generations forward, as the Anthropocene asks us to do, is a true challenge."[30] For the philosopher José Medina, finding such an expansive view of time is possible only with recourse to imagination. As Medina puts it, "Our experiences are expanded and critically assessed in and through the imagination, which enables us to connect our actual experiences with possible ones, extending and projecting them into alternative pasts, presents, and futures."[31] The Indigenous philosopher Kyle Whyte draws out the implications of telling time through deadlines and lifespans.[32] Whyte presents kinship time as a way of noting shifts in relations and therefore taking on the responsibility those relations entail. Kinship time, as elaborated by Whyte and as lived in the

Anishinaabe and Mishipizhu traditions upon which he draws, presses us to augment rhetoric's "art of making things matter" with questions such as: matter for whom and with what consequences, what harms? Grasping the magnitude and complexity of climate change means attending to time as felt, and felt differently, by different communities and individuals. What is needed here is first to hold open and then to connect multiple conceptions of time, tasks that call on all available imaginative resources, including—and especially, I argue—those of rhetoric.

This book's rhetorical approach asks how time gets told by stretching imaginations in rituals, in testimonies, in material, and in sensible encounters, and it suggests the difficulty and the importance of simultaneously holding open multiple kinds, directions, and spans of time. In the fleeting shadow of the occasion, the moment (what ancient Greeks called *kairos*), the other, more widely known conception of time (*chronos*, time's linear unfolding) has receded in importance. And yet the difficulty is this: *chronos*, too, must be held in mind.[33] And not just the *chronos* of clocks and calendars, but that of eras and indeed epochs. In other words, an insistence on time's quality, a too-narrow focus on *kairos*, has perhaps dulled other versions of telling time, versions that of late have come to matter. And this shift foreshadows the other ways that rhetoric has changed and must change, how its work intensifies.

Witnessing

The need for mixed temporalities and new imaginative pathways gives rise to another set of findings involving arts and acts of witnessing, which I conceive capaciously as weighty assertions of material presence that lay bare injustices and demand a reckoning. This definition draws on the vast literature on witnessing, the bulk of which comes out of work on the Holocaust and memory studies, folded together with treatments of witnessing by scholars working in environmental justice and climate rhetoric.[34] When combined, these scholarly frameworks provide insights on witnessing that are further elaborated in this book: 1) it foregrounds justice and morality; 2) it slides back and forth in time; and 3) it relies on material presence.

The first insight about morality and justice is crucial for keeping witnessing front and center in the context of the climate crisis.[35] The second and third insights about temporality and materiality help to elucidate how, specifically, climate change is stretching and altering arts and acts of witnessing. Put another way, witnessing is crucial for climate rhetoric because of its well-established emphasis on morality and justice. But in the context of climate change, temporality reaches ever futureward, and the conception of a witness broadens to include non-humans as well as humans. These two key changes to rhetorics of witnessing are documented in this book's body chapters.[36]

Witnessing's emphasis on morality and justice is implied in, if not presumed by, just about all the work on witnessing discussed or cited herein, but certain scholars make it explicit. This emphasis on just action makes witnessing a more effective framework than, say, narrative or storytelling. In *Commonplace Witnessing*, Bradford Vivian shows how even the most ordinary forms of witnessing bind together moral sentiment and civic action.[37] John Durham Peters attributes witnessing's "extraordinary moral and cultural force" to its "three sources of law, theology, and atrocity."[38] Writing in the specific context of witnessing human trafficking, the scholar-activist Annie Isabel Fukushima finds in witnessing—in contrast to spectating—a call to action.[39] This pull to action is what couples past with future. The whole point of witnessing, traditionally conceived, is to bring the past to the now, ostensibly to improve conditions for current and future generations, in many cases even "speaking," as Vivian puts it, "for the ears of the future."[40]

To be sure, concepts of witnessing developed in the context of memory studies move between past and present and also entail a concern for future generations. Scholars working on matters of climate and environment are beginning to stretch witnessing even further futureward. Shela Sheikh, a critical ecologist, wonders whether or not it is possible to bear witness to "both present and future experiences"—that is, to understand witnessing as "an accumulation of grievances in the context of environmental degradation," or in other words, as an unfolding.[41] Chapters 2, 3, and 5 of this book not only offer affirmative answers to Sheikh's question; they introduce qualitatively new ways to talk about time, whirling together past, present, and future. These conceptions of witnessing make future memory present (chapter 2), bear witness to how an uncertain future feels right now (chapter 3), and place haunting sensory remnants of past presences—dying trees, recorded animal sounds—in the now (chapter 5). Memory and witnessing, in these instances, are dislodged from the past and placed in relation to ongoing violence and future atrocities. They spiral through vast expanses of time, presenting important temporal alternatives to the linear temporalities of apocalypse.[42]

Presence

The third element of this book's conception of witnessing, then, and one of the means of intensification this book draws out, is the idea of physical presence. In the context of witnessing, presence is figured as a kind of bodily, material presence, what the rhetoric and environmental justice scholar Phaedra Pezzullo frames as the tangible, the sensible, the being there or having been there.[43] Its dependency on presence opens up witnessing to nonhumans, including nonhuman animals, plants, and elements. Etymologically, such an expansion is not that much of a stretch in light of a set of valences of witnessing labeled as "figurative" in the *Oxford English Dictionary*: "to furnish evidence or proof of; to be a sign or mark of, betoken."[44] The accompanying textual evidence extends to stones, water, riverbanks, and the sun. This "figurative" meaning, indeed, yields a poetics of material presence, to which Peters attends. Peters puts it somewhat chiasmatically: "witnessing uses words to approximate presence; body-witnessing uses presence to approximate words."[45] Rhetoric, as has been established here and elsewhere, is of course more than words, but the salient point lies with the act of approximation. The "deep assumptions about the physicality of witnessing" that Peters lays out, and the priority those assumptions grant to presence, hold open the possibility of what this book calls elemental witnessing.[46]

To conceive of elements and their composites as witnesses, as this book does in the case of mountains, ice, stones, and trees, is to acknowledge both the material and poetic force of their presence. These entities need to be heeded in ways that include but also exceed the culling of scientific evidence, or worse, extraction for industrial use. More than sources of proxy data, mountains, trees, water, and stones can work as proxy witnesses. To heed them as such is to activate a poetics of witnessing that moves closer to what environmental scholars and activists call "coexistence."[47]

The expansion of witnessing through its reliance on presence includes nonhuman as well as human animals. Elsewhere I have documented the work of nonhuman animals as a certain kind of witness to pain and suffering in the context of fable.[48] But the nonhuman witnessing that is emerging in the context of climate change presents a more fully dimensional creature, a creature in its own habitat and time.[49] Indeed, nonhuman animals as witnesses can reinforce witnessing's multidirectional temporality as seen in the work of the environmental scholar Michelle Bastian, who proposes that humans heed lessons provided by leatherback turtles. These lessons include a concept of deep time and the slowness of humans to respond. Bastian proposes a "turtle clock" as a way to measure a reckoning of habitat destruction.[50] Such heeding is tied to a logic of witness in Anna Boswell's consideration of the tuatara, a lizard-like reptile originating in what is now New Zealand, as "the planet's oldest living witness to how life-in-place has unfolded to date," and as "both observer and casualty of the drastic settler colonial lifeworld reconstruction that has produced climatic upheavals."[51] And in literary studies, Sarah Grieve effects just such a transformation by finding a poetic expansion of witnessing to nonhuman animals in Muriel Rukeyser's *The Book of the Dead*.[52] Such forms of nonhuman witnessing activate just the kind of "potent ethical and political possibilities" that, according to the environmental studies scholar Stacy Alaimo, "emerge from the literal contact zone between human corporeality and more-than-human nature."[53]

Trees, too, are figured as witnesses in a number of contexts, ranging from poetry to settler surveys to war; they offer an enduring example of witnessing's poetic and material resources.[54] In a study of Vaia, a disastrous storm that struck Italy's Fiemme valley in 2018, the environmental anthropologist Nicola Martellozzo observes that "like humans, trees also witnessed [the storm's] impact."[55] Martellozzo's casting of trees as witnesses elevates them from the status of evidence, from a kind of forensic frame that asks "what happened here?" to a moral frame, one that asks about the toll on lives, human and nonhuman. Calling forests "embedded witnesses," that is, acknowledges their presence, their lives, their forms of "telling." This move may tilt toward the figurative sense of witnessing, but it is one that nevertheless works materially, and, as most all witnesses do, by proxy. Elemental witnessing therefore calls forth multisensory listening, moving listening into the more profound, more action-oriented responsibility of heeding.

Such an expansion of witnessing through presence to include animals (human and nonhuman), plants, and elements is therefore one of the important shifts that climate change is bringing about, and one that this book further elaborates in chapters 2 and 5 and in the epilogue. At stake in elemental and nonhuman animal witnessing is the need to find new, more equal, more caring, and more just modes of relation. As Jeffrey Jerome Cohen and Lowell Duckert put it in their return to elements as the basis for ecocriticism, "the less human the collective, the more humane it may become."[56] Instances of elemental witnessing dot this book, and they join together with animals human and nonhuman to constitute capacious arts and acts of witnessing, as what Cohen and Duckert term "partners in material vitality."[57] What results from these new forms of witnessing, as I will show, is the potential to hold open multiple timescales at once, to undermine the peculiar (Western) human habit of thinking in terms of single lifespans. Using presence and temporality to enlarge existing conceptions of witnessing reinforces, in turn, the stable feature of witnessing as an act and as an art: its commitment to justice, its promise for care.

The preceding discussion of witnessing shows how dependent witnessing is on material presence. Presence stands on its own as a robust concept in rhetorical theory, having been a central concept in the field since at least the middle of the twentieth century when Chaim Perelman, writing alone and in collaboration with Lucie Olbrechts-Tyteca, offered it as a core technique of argumentation, a matter of emphasis.[58] As Perelman puts it, "Things present, things near to us in space and time, act directly on our sensibility. The orator's endeavors often consist, however, in bringing to mind things that are not immediately present."[59] In the context of witnessing, as Vivian notes, witnesses stand in for the "radically absent."[60] Rhetoric's task is to conjure presence where there is none. The point of rhetoric, or one important point, is to focus attention on what Perelman calls "certain elements . . . by endowing them, as it were, with a 'presence.'"[61] Presence, or the act of giving presence to "certain elements," lies at the very heart of rhetoric. As the rhetoric scholar Carole Blair puts it, "rhetoric is not rhetoric until it is uttered, written, or otherwise manifested or given presence."[62] This book introduces a new stress on the word "elements" in Perelman's conception, moving it from a general word for "things the rhetor wishes to emphasize" to earthly elements, the material bases of existence.

Presence, then, intensifies by bringing forth the vital matters of accessibility, proximity, and (and through) sense-ability. Given that so much of the climate crisis seems too distant, too difficult to imagine, its unfolding too slow for non-scientists to register, presence takes on even greater importance, and its techniques are reaching deeper into sensuous resources of words, images, art, ceremony, and yes, elements.

Magnitude

So too magnitude. Another of this book's central concepts, the capacious word magnitude names a specific technique of presence that carries and conveys weight and enormity. Magnitude has seen a recent surge in rhetoric scholarship, where it deals with, as Farrell puts it, "the gravity, the enormity, the weightiness of what is enacted, a sense of significance that may be glimpsed and recognized by others."[63] Or, as he puts it in another way and place, and in a passage that is frequently quoted: "magnitude says 'Hey, look at this! This is important!'"[64] The rhetoric scholar Cara Finnegan ascribes to photography a unique ability to confer magnitude—visibility, significance, weight.[65] The point is that magnitude can achieve several effects simultaneously. Working through the senses to convey importance, magnitude is a term of both quantity and quality. As a rhetorical principle, magnitude endows things with size and weight. It can also easily overwhelm.

Christa Olson's treatment of the term and its Greek forebear, *megethos*, locates it "in the aspects of rhetorical practice aimed at community formation, witnessing, and sense of shared place."[66] In the climate crisis, magnitude—size and import—retains its prominent place in arts and acts of witnessing, but in climate rhetoric, magnitude is not always the aggrandizing, bloviating style of assertion that Olson finds in American magnitude.[67] Olson's American magnitude certainly permeates arguments about resource extraction, extreme consumption, and individual rights, a point I return to in chapter 4. But in the approaches examined in this book, magnitude becomes a means of intensification with potential to galvanize action. The magnitude that shows up in chapters 2, 3, and 5 is constituted by sensuous presences and felt temporalities. Chapter 4 links magnitude to mathematical sensibilities, specifically the often confounding concept of exponential growth. Magnitude, in short, is quickly becoming the rhetorical feature that attaches feeling to information, and efforts to convey magnitude in the context of the climate crisis show most pointedly how this so. As Jenny Rice puts it, magnitude names "how abundant information accumulates in ways that expand beyond epistemic registers, creating a sense of coherence."[68] Rice shows how magnitude pulls information into the realm of aesthetics, sensation, and feeling.[69] Magnitude, in the context of the climate crisis, becomes a term of extremity, of eye-widening unease.

The word magnitude appears in the 2021 IPCC report mentioned above (AR6 Climate Change 2021), especially in the eleventh chapter on "Weather and Climate Extreme Events in a Changing Climate," often in close proximity to the words "frequency" and "extremity." Authors of the report use magnitude to present climate change as an intensifier, aligning it with the kind of intensification discussed at the beginning of this introduction. The most familiar geoscientific use of magnitude is as a quantitative measure in the context of earthquakes: the Richter magnitude scale, as is widely known, measures the amount of seismic energy released in a quake. Floods too are measured by magnitude, on the Dartmouth Flood Observatory (DFO) magnitude scale, with the number 10 being reserved for "a flood of record."[70] That phrase is tied to more recognizable characterizations such as, say, a "500-year flood." (More on these labels in chapter 3.) As a rhetorical concept, magnitude gathers into its valences these associations with severity and harm.

Magnitude can sometimes be graphed onto a scale (as with the Richter magnitude scale or the DFO magnitude scale), but magnitude is not identical with scale. Generally, scale provides a way of discussing a graduated series such as a musical scale. Scale is helpful for considering size as it expands and contracts at discernible intervals, as in, for example, Birgit Schneider and Lynda Walsh's exemplary consideration of the "visual downscaling" of images to convert global to regional data.[71] Rhetoric scholars and policy scholars alike discuss the power and rhetoric of what they call "scale framing."[72] When the rhetoric scholar Ned O'Gorman finds an emphasis on scale in Aristotle's conception of epideictic rhetoric, he finds it by placing the act of comparison next to the concept of *megethos* (magnitude), so that size is made relative and turns into scale.[73] As a term of graded quantity, scale is useful for thinking about proportions and relations, and it stands as one of the undisputed communicative obstacles for climate scientists, writers, and activists.[74] I use scale specifically in this book to think about timescales, and I argue in chapter 2 that the amplification in epideictic rhetoric, for example, moves along scales. When DeLoughrey attributes to "enormous scales" a "universalizing geologic," she hints that scale can at times tip into totalizing and even a kind of universalizing.[75]

Magnitude, though, in contrast to scale, encompasses both quantity and quality, and it does so simultaneously.[76] According to the *Oxford English Dictionary*, it names "the quality or fact of being great (in various senses)" and then goes on to specify a "physical sense" and a sense of sound, loudness.[77] The word magnitude tilts in the directions of both sensible greatness (i.e., sheer size) and abstract greatness (honor or gravity). It gathers feeling on multiple registers—it is, in short, a term of intensification. Like witnessing, it carries overtones of severity and even harm such that it can shift according to experience and place, therefore resisting the universalizing tendencies of scale. I have chosen it deliberately.[78]

Temporally expansive acts and arts of witnessing depend on presence to create and enhance magnitude, to activate imagination. These main concepts course through the book, and they move along on a current of other concepts, terms for genres, techniques, time, and feeling—epideictic rhetoric, placement, *kairos*, *chronos*, and rhetorical mood. Few of these concepts are new—indeed, many are quite old—but they are intensified, illuminated anew by the novel situation.[79] Other concepts—like felt time (chapter 3) and mathematical magnitude (chapter 4)—are refinements of existing ones, and I elaborate them with the aid of disciplines ranging from public health to ecological science to mathematics. Together, the cast of concepts featured in this book reveal how rhetoric is shifting in response to the too rapidly changing climate.

Methods, Materials

To theorize the felt rhetorics that climate change is drawing out demands conceptual range, and it also requires methodological flexibility. The suite of methods I use to write this book allows for variation across the different chapters, depending on the materials available in each instance. Rhetoric is itself a nimble, flexible art, and its study demands flexibility as well. The methods therefore shift and mix according to the temporal and material circumstances of the events I study: a ceremony just past; two congressional hearings held two years prior; two crises unfolding in relation to each other, in real time, but at markedly different rates; a large-scale art installation planned, deferred, and then brought to fruition a year later.

All chapters draw on traditional library research, though what that meant during a pandemic, when institutional libraries closed for months and physical materials were not accessible, was a near exclusive reliance on digital resources. For chapter 2, I interviewed and corresponded with key participants and creators of a ceremony for a lost Icelandic glacier. My interlocutors shared documents associated with the ceremony—such as the glacier's ceremonial death certificate, the formal proposal addressed to municipality officials, and photographs of the site selection for a memorial plaque whose placement was part of the ceremony. For chapter 3, I studied congressional transcripts—a well-established approach in political and policy rhetoric—in combination with video recordings of two climate hearings available on YouTube and linked from the two senate committees' web pages.[80] The videos provide access to witnesses' tone, pitch, facial expression, bodily disposition, pace, and energy, all indispensable for the resulting account of felt rhetoric. I also studied publications and social media posts, where available, by the congressional witnesses. Doing so provides further insight into their expressed commitments, their rhetorical strategies, and their lived experiences, all of which constitute their arts and acts of witnessing. Because I wrote chapter 4 during a newly unfolding crisis, news reports, press conferences, and social media posts are front and center in a way that they usually are not in my research, but that is where the story of climate workers' appropriation of the pandemic's curve—soon to be curves—played out. For chapter 5, which focuses on a public art installation, I traveled three times to the site of the installation, participated in a small press session led by the artist on opening day, and observed the material and sensory mixings the installation staged as the seasons changed.

The book's methodological and material range differs somewhat from my previous books, where the methods are more archival and historical.[81] Yet the interpretive strategy in this book nevertheless resembles that which I have used in past books. That strategy can perhaps best be described as an engaged and open following, one that aims to *make sense* of what is happening from the point of view of rhetoric. To "make sense" cuts in two different directions. The first is to understand and clarify. But even those actions can cut away from representational meaning, which adheres too closely to logical certainty. To "make sense," here, is also to engage with, activate, and in some cases speak the senses. To make sense is a matter of deep and sometimes visceral feeling.[82] Rather than viewing these related categories—feeling, emotion, affect—as factors that "short-circuit" rational deliberation, I approach them as indispensable constituents of rhetoric.[83] Feeling and its bodily, sensuous manifestations matter for rhetoric—always have, always will. And rhetorical responses to the climate crisis are illustrating that point with remarkable clarity. And so, as I studied, encountered, and assembled this book's materials, I attended to their glimmers of sensation, for insights they offer on how rhetoric's seemingly durable concepts hold up. They are holding up by doing exactly what rhetoric, as a nimble, contingent, responsive art, does best, which is to change. Rhetoric's concepts, that is, are flexing, stretching, and transforming to meet the imaginative challenges posed by a rapidly changing climate. What this study, this following, brings out, then, is a discernible set of felt rhetorics that exhibit new forms of witnessing.[84]

Chapters: Departures

The book examines four departures from rhetoric's playbook in the context of the climate crisis: a memorial to a glacier; congressional testimonies of youth climate rhetors; the borrowing of visual and mathematical resources from another crisis; and a ghostly public art installation. These departures work something like "case studies," granular examinations of a particular event or phenomenon, with the aim of identifying what is new here.[85] These departures present work-arounds, which is to say new resources and different means of intensification, in response to rhetorical situations with few or no precedents.

The departures all differ from each other. Such variety is a deliberate choice on my part.[86] Each chapter inhabits a particular "sphere of resonance" for scholars of rhetoric and communication, and, I hope, for anyone interested in how people are urging each other to care about and do something about the changing climate.[87] Those spheres—memory studies, public address, public health, public art—are themselves gathering intensity as the climate crisis worsens. Each chapter spotlights a set of climate arguments that works at the level of imagination more intensely than it does at the level of logical argumentation or information sharing. Each one uses techniques of witnessing meant to pull people in. Rather than aim for comprehensiveness, a goal that is elusive if not illusive, the book's plan and my choice of departures favor vividness and suggestiveness. Together, the chapters work perhaps like a prism, where the chapter-angles refract concepts to create flashes of distinctive clarity.

The chapters show witnessing sliding back and forth in time, and they do so by moving chiasmatically from memory to uncertain futures back to memory again, all by way of haunting. Chapter 2, "Glacial Death: Making Future Memory Present," examines a memorial ceremony held in 2019 for an Icelandic glacier that recently lost its status as a glacier. That ceremony uses age-old features of epideictic rhetoric even as it alters them. Building on an interview I conducted with Howe and on other publicized interviews and writings by Boyer, Howe, and the Icelandic writer Andri Snær Magnason, who served as something of an emcee for the ceremony, the chapter renovates Mikhail Bakhtin's concept of the chronotope, which names a verbal-sensory manifestation of time and place. A set of specific chronotopes—wishes, glaciers, generations—allowed organizers and speakers at the event to exploit the temporal and generic flexibility of the epideictic genre. In this way, the ceremony shows how chronotopes help audiences vividly imagine a future in the assigned time of the present—how chronotopes can, in effect, make magnitude of ice loss more accessible to those for whom ice has little or no cultural weight. At the heart of the investigation is not just the fact or act of attention, central to the art of rhetoric, but the quality of that attention—the sensory, felt dimensions of the ceremony.

Chapter 3, "In a World Full of 'Ifs': The Felt Time of Youth Climate Rhetors," turns from the temporal malleability of ceremonial rhetoric, how it can move back and forth in time, projecting the present into the future as an imagined past, to what time feels like to those staring down a grim future. This chapter, that is, draws a rhetoric of felt time from youth climate rhetors, especially BIPOC youth climate spokespeople who live in the United States. It does so by focusing on what a handful of such young people said when invited to testify before Congress in two separate hearings, both held in 2019. These testimonies merge time and feeling, two broad categories that have long concerned scholars and theorists of rhetoric. The result is a rhetoric of felt time, a sense of urgency that stresses the word "sense" anew in this otherwise common-place phrase. The chapter identifies the features of felt time, how these rhetors in particular describe and live it—that is, bear witness to it—and how they try to help older generations of lawmakers feel it with an urgency approximating their own.

Chapter 4, "Learning Curves: COVID-19, Climate Change, and Mathematical Magnitude," returns to chapter 2's treatment of magnitude, this time to detail how one unprecedented crisis (the COVID-19 pandemic) made a new rhetorical pathway available to another (the climate crisis). This chapter examines the early rhetorical life of the "flatten the curve" campaign and its data visualization, with particular attention to the quick, fleeting leap the slogan and its defining image made into climate discourse. A careful consideration of that leap reveals a dimension of rhetorical magnitude as yet unconsidered: that of mathematical magnitude. Specifically, the chapter shows how these curves depict a mathematical magnitude that takes on and carries feeling from one unprecedented crisis to another. Following the circulation of this graphic and its phrase offers a way to document rhetorical effects.

In presenting two futures for comparison at the beginning of the pandemic, one steep curve and one flatter one, the image projects the effect of mitigation. Mitigation in the case of the pandemic meant a combination of policy and campaigns for individual behaviors (e.g., physical distancing, mask use, the limiting of large gatherings, and the like). Mitigation in the case of the climate crisis means a different set of policies whose effects, like the pandemic, nevertheless can be reproduced comparatively (e.g., graphed in side-by-side curves that show what will happen with and without mitigation). The image—and, as the chapter argues, the phrase "flatten the curve"—quickly conveyed the importance of mitigation measures, while also conveying the astounding magnitude of exponential growth and carrying with them the complex and conflicting feelings of the pandemic.

Chapter 5, "Presence and Placement in Maya Lin's *Ghost Forest*," focuses on a temporary public art installation that the well-known artist Maya Lin created by hauling forty-nine nearly dead Atlantic white cedars from the Pine Barrens in New Jersey and arranging them into a stand in the middle of Madison Square Park in Manhattan. The trees were dying because of over-salinization from rising seawater and climate-intensified storm surges. Their arrangement in a lush, season-shifting urban park, itself classified as an arboretum, presents an opportunity to study anew presence, a concept long considered important to the work of rhetoric, and placement, a concept that holds possibility for reducing distance.

An epilogue, "Fathoming," serves as a conclusive yet open-ended, and perhaps hopeful, capstone to the book and returns to the ceremonial overtones sounded by the genre of epideictic rhetoric considered in chapter 2, as well as to the matter of trees brought to this study by Lin's *Ghost Forest*. This epilogue, though, meditates briefly on the veneration of trees by artists, writers, and speakers with a focus on Indigenous teachings on the matter. The chapter offers fathoming and veneration as crucial rhetorical acts, reminders of how water, land, and trees have long been considered sacred by Indigenous people worldwide. Trees in particular, whose lifespans often far exceed those of humans, can, through their elemental witnessing, help meet the challenges of scale, to correct the too-narrow, too-Western focus on individual, human lifespans.

All told, this study offers an *in situ* account of novel rhetorical theory and practice, elaborating the concepts and approaches that are turning up in this wickedly complex situation that is indeed without precedent.[88] That account focuses on the means of intensification—the visual, the visceral, the imaginative—the means, that is, that stretch beyond, but nevertheless still at times entail, reason. In the climate crisis, these means of intensification call upon newly enlarged arts of witnessing, magnitude, and presence, all of which urge a relationship to time that is at once capacious, nimble, and felt. These means of intensification, that is, encourage ways of imagining and planning futures without the assistance of precedent. Just as the frontispiece of this book captures a moment of intensification by means of elemental, particulate, atmospheric, and optical conditions, this book finds moments when rhetoric works vividly and profoundly as an art of intensification, an art that travels through and alters feeling and sensation as much as, and often in conjunction with, information and argument. As an art, rhetoric can work at some times directly, at others more obliquely, and the climate crisis, by altering the conditions of life itself, is calling on rhetoric's capacity to intensify, even as it deepens that very capacity, shaping what is—what can be—seen, felt, and imagined.

CHAPTER 2

Glacial Death

MAKING FUTURE MEMORY PRESENT

For a glacier to be a glacier, ice must continually move underneath the weight of accumulated and compacted snow. That definition no longer applied to Okjökull, a glacier, or former glacier, atop a shield volcano in western Iceland. Its ice had stopped moving. Rising temperatures from climate change had, over time, caused the glacier to recede to a brittle, glazed lake dotting the top of the volcanic mountain known as Ok. (In Icelandic, "jökull" means glacier, so the area is now referred to by the mountain's name, Ok, pronounced "awk.") The mass of moving ice that once spread across six square miles in 1890 had dwindled to a less than half a square mile patch of what glaciologists call "dead ice."[1] You can see the glacier's recession in aerial photographs of the area from 1986 (figure 2.1) and 2019 (figure 2.2), circulated on social media by NASA on August 18, 2019.[2]

The Icelandic glaciologist Oddur Sigurðsson declared Okjökull dead in 2014.[3] On August 18, 2019, Sigurðsson boarded a bus in Reykjavik with a death certificate for Okjökull in his backpack, which he had signed earlier that day (figure 2.3). The ceremonial death certificate lists the cause of death as excessive summer heat, and for the part of the certificate requesting information on "rescue operation," Sigurðsson wrote: "Ekkert var gert til bjargar"—"nothing was done to save."[4] On the bus Sigurðsson was joined by three anthropologists, professors Cymene Howe and Dominic Boyer and graduate student Magnús Örn Sigurðsson, all from Rice University in Houston, Texas; the Icelandic writer Andri Snær Magnason; and dozens of climate activists, including several children. The bus headed northeast, to Ok's base, where the group was joined by Iceland's prime minister, Katrín Jakobsdóttir; Mary Robinson, formerly the UN High Commissioner for Human Rights

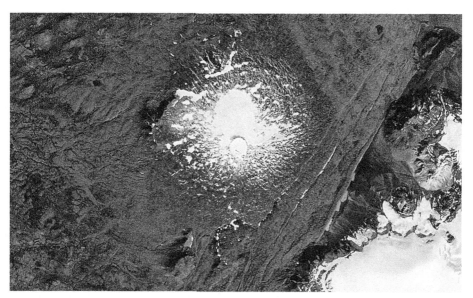

2.1 Okjökull in 1986. Courtesy of NASA Earth Observatory. Image credit: Joshua Stevens.

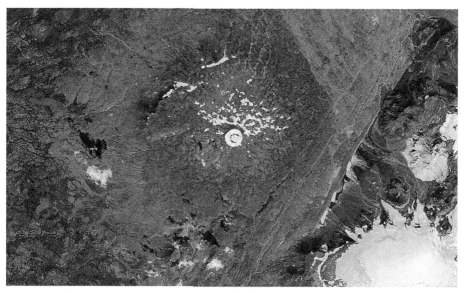

2.2 Okjökull in August 2019. Courtesy of NASA Earth Observatory. Image credit: Joshua Stevens.

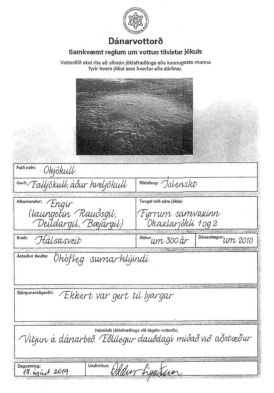

Dánarvottorð

Samkvæmt reglum um vottun tilvistar jökuls

Vottorðið skal rita að yfirsýn jöklafræðings eða kunnugustu manna
fyrir hvern jökul sem hverfur eða stirðnar.

Fullt nafn: *Okjökull*	
Gerð: *Falljökull, áður hveljökull*	Ríkisfang: *Íslenskt*
Afkomendur: *Engir (laungetin Rauðsgil, Deildargil, Bæjargil)*	Tengsl (við aðra jökla): *Fyrrum samvaxinn Okaxlarjökli 1 og 2*
Sveit: *Hálsasveit*	Aldur: *um 300 ár* Dánardagur: *um 2010*
Ástæður dauða: *Óhófleg sumarhlýindi*	
Björgunaraðgerðir: *Ekkert var gert til bjargar*	
Heimildir jöklafræðings við útgáfu vottorðs: *Vitjun á dánarbeð. Eðlilegur dauðdagi miðað við aðstæður*	
Dagsetning: *18. ágúst 2019*	Undirritun: *Oddur Sigurðsson*

2.3 Death certificate for Okjökull. Image credit: Oddur Sigurðsson; design: Aðalsteinn Svanur Sigfússon; calligraphy: Sigurlaug Hjaltadóttir.

(and a past president of Ireland); and Guðmundur Ingi Guðbrandsson, Iceland's Minister for the Environment and Natural Resources. The prime minister delivered brief remarks and read a stanza from a poem written by a youth environmental activist and her father, both of whom were also present that day. Then the group hiked up through the lava fields to the place where Okjökull used to be. There, they intended to hold a ceremony for the glacier, planned and organized by Howe and Boyer, and install a memorial plaque. Just before the group arrived at the top of the mountain, a few hundred yards from where the memorial would be installed, the writer Andri Magnason addressed them. According to Howe, Magnason said that "there's never been a ceremony for a fallen glacier . . . we are going to invent it as we go along. We are going to have to create it as we go."[5]

The unprecedented nature of the ceremony and memorial plaque, and of the loss they commemorate, calls for careful consideration of how they work rhetorically, how they attract and hold attention, and how, specifically, they activate imaginations in order to craft public sentiment.[6] Of particular interest to environmental communication scholars and theorists of rhetoric more generally is how the ceremony and memorial bring distant time and place close; how they enlarge what Howe calls "a small story," how they give that story magnitude.[7] To elucidate the distinctly rhetorical work accomplished by the ceremony and the memorial, I offer a renovated version of the chronotope, that concept binding time (*chronos*) and place (*topos*) put forth almost a century ago by the Russian philosopher and literary theorist Mikhail Bakhtin.[8]

The chronotopes layered into the ceremony and its memorial create mixed modes of elemental and human witnessing that turn attention futureward. Chronotopes show forth—they bring before the eyes—time and place by enlivening them, making them accessible, and injecting them with feeling. What is more, the ceremony and its memorial encourage participants and readers to hold open multiple timescales at once, a difficult yet necessary task for facing climate change. Holding open multiple timescales simultaneously is a feat accomplished in, through, and with imagination, with words activating images, wrapping those images in feeling, and reaching toward distant futures and places. Imagining is how one sees past the vanishing point extending from the now or near now, and it therefore offers a theoretical approach that is slightly different from but complementary to the existing work on perceptual difficulties posed by climate change.[9] And chronotopes, specific time/place focal points, help in particular to stretch what John Durham Peters calls our "scalar imagination," which is often limited like, if not by, sensory experience.[10]

With scalar imagination at the fore, the first part of the chapter will consider epideictic rhetoric—the honorary, ceremonial genre—as particularly fertile ground for the chronotope's witnessing work. It will then present a full rhetorical account of the chronotope, considering how it operates as a means of witness in the roomy genre of epideictic rhetoric. The ceremony on Ok stimulates imaginative actions with three chronotopes: wishes, glaciers, and generations. The first chronotope (wishes) is as wispy, light, and fleeting as the second (glaciers) is weighty and enduring. One nudges attention to a future not yet witnessed, whereas the other works with time as sequestered in the physical place of the mountain, the former glacier, Okjökull. The third chronotope works with—and across—generations, conjuring an idea of ancestorhood that both draws on and inverts the ways ancestors are often imagined. The generational chronotope, in this context, projects both ancestry and witnessing forward rather than backward in time.[11]

The second part of the chapter considers the memorial itself, its materiality, and the imagined public memory it conjures with its curious chronotopic form, "a letter to the future," the text for which was composed by Magnason. The material chronotope of the rock to which the memorial is attached, together with the chronotopic image of future readers standing in front of the rock, infuse Magnason's words with, and hold promise for, a future public memory of human action. The remote memorial site, that is, conjures actions that may be remembered at a remote point in the future.

Remoteness—of geographic location and of future consequences—tends to blunt the force of that which is unprecedented. That is why the *stuff* of rhetoric—words, images, data—and the ability to help rhetoric's material travel, circulate, and bring near that which is far, matter all the more. At stake in the analysis is the work of making the unfathomable fathomable by making it accessible, palpable, seeable, *felt*—by, that is, bringing the distant close. The act of making something fathomable can do more than draw attention to that something; it can shape the very quality of that attention. In this way it holds promise of answering Lawrence Buell's call for a "reorientation of human attention and values according to a stronger ethic of care for the nonhuman environment." Such a reorientation, Buell contends, "would make the world a better place, for humans as well as for nonhumans."[12] By connecting and moving images of time/place, and by shaping the quality of attention given to an issue as complex as climate change, chronotopes can meet the distinctly rhetorical challenges of imagining futures without precedent—futures both close (the urgent time for action) and distant (the time when the effects of any action or inaction will come to pass). For in the context of the climate crisis, futures near and deep are, finally, what need to be made to matter.

Chronotopes, Rhetoric, Attention

The importance of time, timeframes, and timescales gives chronotopes new salience in the context of the climate crisis. So what, exactly, are chronotopes for Bakhtin? For all his fondness for the concept, Bakhtin refrains from offering a single definition, opting instead to describe the work it does, particularly in novels. Toward the end of his long essay, "Forms of Time and of the Chronotope in the Novel," he has this to say: "the chronotope, functioning as the primary means for materializing time in space, emerges as a center for concretizing representation, as a force giving body to the entire novel. All the novel's abstract elements—philosophical and social generalizations, ideas, analyses of cause and effect—gravitate toward the chronotope and through it take on flesh and blood, permitting the imaging power of art to do its work."[13]

The gravitational pull of the chronotope can be seen in the chronotope of the road, one to which Bakhtin ascribes "immense" importance. In novels, that is, the road chronotope facilitates meetings that happen in what Bakhtin calls "chance time," a kind of time that injects an immediacy and even a suddenness into a novel's movement.[14] The 2014 novel *Station Eleven*, as one example among many, makes use of the road chronotope, as a traveling band of performers follows mostly abandoned roads and pathways through a desolate landscape years after a deadly pandemic swept across the earth. The routes and paths the troop follows are themselves the source of drama and surprise, a place of encountering strangers—a child sentry here, a stowaway from an earlier stop there; the road chronotope heightens the unpredictability of chance time.[15] Late in the essay Bakhtin returns to the road, describing its chronotopic work as follows:

> The road is a particularly good place for random encounters. On the road (the "'high' road"), the spatial and temporal paths of the most varied people—representatives of all social classes, estates, religions, nationalities, ages—intersect at one spatial and temporal point. People who are normally kept separate by social and spatial distance can accidentally meet; any contrast may crop up, the most various fates may collide and interweave with one another. On the road the spatial and temporal series defining human fates and lives combine with one another in distinctive ways, even as they become more complex and more concrete by the collapse of *social distances*.[16]

The road helpfully illustrates how, with chronotopes, "time and space fuse together," how a literary image comes to be more than an image: it makes worlds, organizes experiences, and facilitates actions.[17]

These world-making capacities make the chronotope a particularly potent concept for rhetorical theory and criticism.[18] In novels, what matters is the movement of plot and character, but what matters in rhetoric is the movement of attention and action. What literary and rhetorical domains share, though, are language's poetic resources; literary and rhetorical genres both tap into activities of sense-based imagining to show, to show forth, to make worlds, and importantly, to draw audiences into those worlds. Chronotopes provide vivid, concrete "moments of poesis," to borrow an arresting phrase used by the anthropologist Kathleen Stewart.[19]

In addition to the sensory, world-making quality of chronotopes, and their ability to slide imaginative attention back and forth in time, I wish to stress their mutual inclusivity, their connectivity to each other. Jodie Nicotra and Judith Parrish discuss how multiple chronotopes compete and jostle for dominance, and such a view is part of Bakhtin's picture—but only part. For Bakhtin, "chronotopes are mutually inclusive, they coexist, they may be interwoven with, replace or oppose one another, contradict one another or find themselves in ever more complex inter-relationships."[20] Chronotopes, that is, are roomy things, and they can connect one to another. The anthropologist Alaina Lemon characterizes the connective work of chronotopes as layering; Paul Prior and Jody Shipka call it lamination.[21]

The point here is that chronotopes can reinforce, include, and even entail one another. So, for example, in plot movement, Bakhtin's chronotope of the road joins with—it makes way for—chronotopes of meetings and encounters. In rhetoric, this inclusive feature of chronotopes promises what might well be the most important imaginative habit to cultivate in the context of climate rhetoric: the ability to hold open multiple timescales, and to do so simultaneously, in the present. We humans must get better at imagining a deep future, a medium term, and a very near term, to keep them all open and link them together, and to do so at the same time, in the now.[22]

Chronotopes are indispensable to rhetoric's attentional work. For all the existing work on rhetoric's importance for "getting attention," rhetoric scholars would do well to pay more attention to the quality of that attention, the need for it to cultivate care.[23] Chronotopes shape attention's quality, deepen its intensity, by gathering attentive energy around their "knot" of time and place.[24] As Bakhtin put it when he first introduced the chronotope, "temporal and spatial determinations are inseparable from one another, and always colored by emotions and values."[25] The feltness of space and time is exactly what lends chronotopes their intensity in the context of rhetoric.[26] As means of intensification, chronotopes help words to, effectively, sink in, to burrow and nag. Their gravitational pull allows attentional energies to *cohere* and keeps them from scattering.

As such, chronotopes may be defined for rhetoric as sensuous bits of language that materialize time in space and enliven and gather attention's energy. As my analysis will show, chronotopes can carry feeling, thereby shaping the *quality* of attention—deepening it, intensifying it. Put another way, time and place cannot be felt without being fathomed, and that fathoming is the work of imagination aided by the specific world-making capacities of chronotopes.

Epideictic Rhetoric, Chronotopic Witnessing

When imported from literature into rhetoric, chronotopes retain their usefulness for considering genre, specifically epideictic contexts, such as the memorial for Okjökull. Here, chronotopes facilitate witnessing, which, again, this book conceives as weighty assertions of material presence that expose injustices and demand a reckoning. Vivian presents witnessing as a specifically contemporary form of public rhetorical memory, "an especially prevalent, accessible, and affecting mode of address in modern Western societies."[27] While no genre of rhetoric is particularly well equipped (for Aristotle at least) to handle that which is unprecedented, epideictic, the province of ceremonial rhetoric, may still be the most promising, as the anthropologists Boyer and Howe, experts in ritual, seem aware.[28]

Epideictic's foremost tasks are first, to render explicit something already known, and then to intensify preexisting commitments. As Laurent Pernot shows, the work of epideictic is primarily documentary and is carried out in the argumentative form of amplification, which generally entails making certain dimensions of a subject larger or louder.[29] What Pernot has to say about amplification in epideictic is critical for an understanding of the ceremonial genre and how it works rhetorically: "amplification does not mean 'development,' even less 'padding out.' It involves not lengthening a speech but increasing the size of its subject, by emphasizing its beauty, its noblesse, etc." Neither, Pernot continues, is amplification a matter of "exterior ornament."[30]

Amplification moves along scales. It presents images before the eyes of the listeners, the public, and it magnifies their importance, thereby shaping attention's quality and turning into magnitude. Amplification of the circumstances of Okjökull's disappearance stands as the primary purpose of the ceremony and memorial on Ok. As Howe put it to me in our conversation, "the whole reason we wanted to put [together] the memorial in the first place was really to, I don't know, it's not exactly to draw attention to climate change because everyone knows what's happening, but as a spur to action to recognize that [knowledge]."[31] Howe's comment here seems to intuit what tips attention into action—the recognition and acknowledgment that constitutes witnessing, that gives it weight. The remarks Howe composed for the ceremony illuminate the day's purpose for the group: "Let us remember that a memorial, like this one, is to commemorate something that has been lost.

But memorials are, in the end, primarily for the living. That is us. And so ultimately this memorial is a reckoning. And a call to action. For us, the living."[32]

For Howe and Boyer, the memorial would document what the poetry scholar Lynn Keller calls "the self-conscious Anthropocene," would grapple with the fact of the widespread awareness of climate change's effects next to insufficient action worldwide. According to Keller, "The designation 'self-conscious Anthropocene' enables us to name this period of *changed recognition* when the responsibility humans bear for the condition of the planet and for the fates of Holocene species is widely understood."[33] The "self-conscious Anthropocene" is largely a contemporary phenomenon, according to Keller, and it calls forth "a cultural reality more than a scientific one."[34] Keller's intervention positions contemporary North American poems about the environment as both witnesses to and resources for encountering the self-conscious Anthropocene. Much as the chronotope easily moves from literature to rhetoric, Keller's concept finds salience in rhetorical genres, again, because both draw on language's poetic—which is to say world-making—resources.

The ceremony on Ok effectively channeled the epideictic orator, who, according to Pernot, "is on a mission, and [whose] speech is a political and social act."[35] And the purpose of reckoning gives the ceremony a moral dimension too, bends it toward witnessing. The documentary work endemic to the epideictic genre, in short, serves the rhetorical purpose of witnessing.[36] As Peters puts it in his reflection on witnessing, "to witness is to wish that the record of the past were more whole, and to grasp this lesson now is to live vigilantly, to make the present worthy as we imagine contemplating it from a future point."[37] In fact, the importance of vigilant witnessing is exactly why Keller's articulation of the self-conscious Anthropocene matters so much to environmental communication and rhetoric. Rhetorically, then, the reckoning on Ok, its insistence that humans acknowledge—bear witness to—what they know, documents, magnifies, and amplifies.

Witnessing's wish for a worthy present as contemplated in the future can be served by chronotopes too, but only if those chronotopes can swivel in another temporal direction. Bakhtin, after all, developed the concept in service of a "historical poetics," and one commentator even glosses chronotopes, with an eloquent flair, as "invokable histories."[38] While the idea of invokable histories—on historical and even geological timescales—inheres in the chronotopes I will examine, chronotopes can nevertheless be positioned as invokable witnesses, sliding back and forth in time, across ever widening timescales, turning attention toward the future. The work of amplification and magnification so central to ceremony—and to epideictic witnessing work—is carried out by means of chronotopes.

So now, equipped with a sense of the Ok ceremony's place within the epideictic genre, the rhetorical genre with perhaps the most flexible temporality, the genre that seeks to bear witness through amplification and magnification by means of chronotopes that are mutually inclusive, themselves sliding back and forth in time, and that are, above all, palpable, *felt*, I wish to turn to the specific chronotopes in the Ok ceremony. I begin with the preparatory chronotope of the wish.

Mood: Wishes

As the group of ceremony participants neared the place where the memorial plaque would be installed, Magnason gathered the hikers together and asked them to finish the hike in complete silence, walking the last hundred meters or so in a single-file line while looking straight ahead. Magnason's instructions were modeled on an Icelandic folk tradition named for a mountain called Helgafell (Icelandic: "Sacred Mountain"), which is located on the Snæfellsnes Peninsula in the western part of Iceland. On the final stretch, the ceremony participants were instructed to "hold good in your heart" and to think good thoughts. If they moved their thoughts toward the good, and if they did so while training their eyes straight ahead, Magnason told them, when they reached the top, they would be granted three wishes. After Magnason explained the ritual, Howe then reflected briefly on the usefulness of wishes in imagining "the future we hope to see."[39] Howe's framing of wishes prompted one of the participants, the writer Lacy Johnson, to reflect on language and scale: "As we walked the last few hundred feet, I realized that we lack metaphors for comprehending the future, much less the scale of the disaster that it has in store for us."[40] Johnson's words pinpoint the need to activate the scalar imagination.

In its small way, the ritual of wish-making addresses the imaginative obstacles posed by the scale of unprecedented loss. The silent contemplation of hoped-for futures prepared participants for the ceremony on top of Ok, with the wishes themselves working as tiny, airy ventures into imagining something better, exercises in hopeful envisioning, no matter how unreal.[41] Wishes work chronotopically by carrying a mood, a grammatical mood, the subjunctive, that has been thought by some linguists to be slipping away from the English language.[42] In particular, wishes fall into a category called "were-subjunctive" or the past subjunctive: "I wish I were x." A wish's rhetorical weight lies in its suggestiveness. A wish encourages the wisher to engage a future imaginatively, and with reference to their individual hopes—to picture what they want to have happen.[43] The subjunctive's winding together of mood and rhetoric is useful to consider because of the mood's nudge toward the future, or an alternative situation—what linguists call an *irrealis mood*, that which is not real (or not yet or not known for certain to be or to have been real), as opposed to the *realis* mood of, say, an emphatic declaration. The irrealis mood, in other words, grammatically demarcates the world of "what might be" from the world of "what is." The subjunctive, then, posits or strains toward a provisional, hoped-for, or tenuously hypothetical situation. The tense flexibility of this mood allows it to cope with multiple timeframes at once.[44] The subjunctive mood is, as a number of public memory scholars have noted, the *mood of anticipation*, as exemplified in Barbie Zelizer's discussion of the subjunctive work of visual images, moving photographs in particular from the "as is" to the "as if."[45] Importantly, in the context of ritual, the subjunctive mood constitutes a shared reality, linking, as Bronislaw Szerszynski puts it, "what is and what could (or ought to) be the case."[46] So in the ritual, ceremonial context, the subjunctive posits a different world, a shared and shareable world, one that, as Clifford Geertz long ago pointed out, fuses worlds lived and imagined.[47]

Once at the top of the mountain, Boyer addressed the group: "We are making wishes today, wishes that we hope will bend the future in our favor."[48] He then shared his wishful vision:

> I live far away from Iceland in Houston, Texas. Houston has been a center of the global fossil fuel industry for almost a century. Believe me that there is nothing that industry and its political allies like to see more than all of us climate types feeling depressed and beaten down. If there's no emotional energy with which to imagine a post-carbon world then of course we will just continue to stare at the slow-motion collapse. So my wish is this: let us be present in our sadness but also for the joy and togetherness that renew us and allow us to thrive.

There's a double mood at work in Boyer's remarks—the first, grammatical, with the subjunctive mood acting as the mood for wishes, for projecting an as yet uncertain future. The second mood, this one more of an affective disposition, follows from the idea of emotional energy that Boyer wants to find in the collective act of mourning. That emotional energy speaks directly to Joshua Trey Barnett's reflective curiosity, which arises in his discussion of ecological grief, about "what capacitates us to feel that someone or something important, meaningful, and valuable has been or could be lost, and that this loss deserves (or would deserve) to be mourned."[49] Boyer's aspiration "to bend the future" toward ecological recovery follows an arc similar to that articulated by the editors of the volume *Mourning Nature*, who "extend the work of mourning to nonhumans to think about other possible futures."[50]

As a chronotope, the act of wishing effectively prepares participants, shaping the quality and intensity of their attention, moving their imaginations futureward while encouraging positive, hopeful thoughts about that future. Boyer's sharing of his wish strove to bind conviction with care. The ceremony and the memorial, that is, make a point, and also sharpen that point with feeling—a kind of sense-enlivened compassion, what Boyer in the ceremony called "emotional energy"—to move attention into action. Therein lies the rhetorical power and flexibility of wishes as chronotopes: their irrealis state can carry regret—I wish certain nations weren't so dependent on fossil fuels—but when trained directly on a changeable future, they can also carry hope. Once participants arrived at the summit, even as they were making their wishes, the chronotope of the wish layered with the ceremony's central chronotope, the very reason for the memorial itself, and the one with the most weight: that of the glacier.

Deixis and Glacial Loss

Glaciers are, of course, the lead chronotope for the ceremony held on Ok—the chronotope that needs little conjuring for the people who walked together up the mountain, many of whom had lived in Iceland since birth. The formal proposal Howe and Boyer sent to Ragnar Frank, the director of Planning and Environment in Borgarbyggð, the municipality where Ok mountain is located, showcases the glacial chronotope, especially in this passage:

> Given the importance of glaciers in Icelandic history, glacial loss is not only an important environmental transformation but also a matter of cultural heritage. Since Iceland has the largest glacial mass in Europe and is well-known internationally for the beauty of its glaciers, this is also a loss that will be felt across the world. In this respect, the fate of Okjökull is a sentinel for the coming impact of climate change and we believe that drawing attention to it will help make both Icelanders and foreigners more aware of the stakes of the struggle against global warming.[51]

The glacier chronotope works in a number of ways in this paragraph. First, glaciers stand as the distinguishing topographical characteristic

of Iceland, and as such they hold the island nation's very identity, its invokable legacy. The presentation of Okjökull's disappearance as a "sentinel" goes further, however, positioning the glacier as a figure of watch-
fulness, a witness. So the ceremony itself honors the loss of a witness.

The proposal, too, mentions the importance of drawing attention to
Okjökull, but what matters, and what the glacier chronotope allows,
again, is the careful crafting of a particular attentional quality. In the
case of Howe's framing for the ceremony, the quality of attention is
one of deep respect and honor, attention that acknowledges the importance of place, of that particular place, the place being honored. Howe's
planned remarks begin with gratitude, to "Ok mountain for having us,
for bearing us, if you will. Ok has been here long before any of us. And
it will be here far longer. I want to respect that presence beneath our
feet, surrounding us today."[52] These opening words seek not merely to
focus attention, but to gather into that attention energy and care for the
land that supports the bodies of participants and their kin, the ground
that used to bear the glacier being memorialized. The chronotope of the
mountain here helps the patch of shared place and the ceremony's tiny
blip of time unfurl backward and forward, the mountain itself standing as a palpably magisterial geological formation, invested with significance by Howe's words at that moment. Howe's planned remarks also
invoke the meaning of the word "ok" in Icelandic—yoke or burden. The
idea of the mountain as bearing or enduring both the ceremony and
humans more generally feeds a poem titled "Ok—The Burdened Glacier," written for the occasion by an Icelandic youth, Gunnhildur Fríða
Hallgrímsdóttir, and her father, Hallgrímur Óskarsson. Hallgrímsdóttir
read the poem aloud before the group disembarked from the bus at the
foot of the mountain.[53]

So the mountain of Ok, its name's poetic valence, and the recent loss
of its glacier weighed on the ceremony and most likely on the participants. Glaciers, of course, more generally, bear cultural, imaginative,
and measurable weight. Only within the last century or so did the word
glacial, as an adjective, acquire the figurative temporal valence characterizing a creeping pace; prior to that, the word mainly meant icy or
freezing.[54] The OED quotations supporting meanings of "extremely
slow" or "very gradually" range from James Joyce's 1922 Ulysses, in
which a "dwarf tree" is described as possessing a "glacial arborescence,"
to a more recent—and indeed more choice—quotation from a New Scientist report on global warming discussions in Washington, DC, during
the early part of 1991: "The US was mainly to blame for the 'glacial pace'
of the negotiations, according to the Climate Action Network for the
US."[55] Indeed, that exceedingly slow temporality is best captured in—

and perhaps even owes to—early methods of confirming a glacier's very status: early glaciologists would place material objects on the ice and then return later (sometimes much later) to see if the objects had moved.

Atmospherically, glaciers form an important line of planetary defense against the sun's intense heat, for snow has the highest albedo, a thermodynamic term used to characterize a surface's ability to reflect solar energy (the word "albedo" derives from *albus*, the Latin word for white).[56] And as the environmental historian Mark Carey demonstrates, glaciers have become "key icons" for the effects of global warming, in part because of the long fascination with them, a fascination that Carey documents and divides into a number of different kinds of narratives. Glaciers work variously in these narratives as natural hazards, as sublime scenery, as scientific laboratories, as tourist meccas, as "remote, empty spaces to conquer," and, of course, as wilderness.[57] "Glaciers have long ignited people's imagination," writes Carey, and the rhetorical power of the glacier chronotope, then, is that it offers a kind of imaginative flexibility, such that images of glaciers and their melting are easily conjured.[58] As Carey argues, glaciers have of late come to be figured as endangered species, and this is precisely because they are thought to be living. The same logic, of course, undergirds both the death certificate and the memorial ceremony created for Okjökull.

Ok bears the glacial chronotope through the remains of its former glacier—in the snow patches that dot the volcano and in the crater lake visible in the NASA photograph. But Ok's glacial neighbors loom with a haunting and visible presence alongside the remains of what was once Okjökull. The surrounding glaciers stand as gleaming reminders of potential future losses, the losses to which Ok serves as a sentinel, to return to the language of the ceremony's proposal. In our conversation about the ceremony's design, Howe described to me the trek she and Boyer led to Ok's summit the year prior, in August 2018, for an event they called an "Unglacier Tour," designed to counter the glacier tourism that has exploded in Iceland in recent years.[59] On this tour, the group members were also scouting a spot for the memorial plaque. Howe described the scene: "we hiked up, we went all the way up to the crater of Ok where you can get the best view; it's super spectacular, you can see all these other glaciers from there, another shield volcano, and so we had beautiful photographs that people took from that."[60]

The surrounding glaciers, Geitlandsjökull, Þórisjökull, and Eiríks-jökull, exert a majestic rhetorical power and carry intensities that vary according to viewers; for visitors they are often considered sublime. Glaciers have long carved the cirques and fjords of Iceland's distinctive landscape.[61] For centuries Icelanders have lived alongside glaciers, farming the soil enriched by their meltwater. For the people of the highlands, these glaciers form the long horizon. One such person, Guðmundur Ingi Guðbrandsson, now Iceland's Minister of the Environment, offered these remarks about the horizon during the ceremony on Ok: "'as a little boy not very far away from here, my grandmother taught me the names of all the mountains we could see on the horizon, and the names of the four glaciers,' Guðbrandsson said. 'When I visit my parents today on their farm, I can see only three.'"[62] Guðbrandsson's remarks present two chronotopes—that of the glacier, and that of familial generations, both wound together by the grandmother's teaching. The loss, for Guðbrandsson, is felt by visual subtraction: four minus one. Guðbrandsson's words, together with the location of the ceremony itself, direct the eyes toward a changing horizon, the visibly vanishing cryosphere. The larger surrounding glaciers have, by most estimates, another century or century and a half before they too recede, like Okjökull, into nothingness.[63]

As such, the surrounding glaciers exert a strong deictic pull. The ancient Greek word *deixis*, demonstration or reference, roots the word "epideictic" and derives from the verb *deiknumi*, to bring to light, to show forth, to display. In English, deixis is used most commonly in the context of linguistics to refer to the basic act of referring, which can occur through words such as "here" or "there," or in the case of temporal deixis, now or then. Deixis can also be accomplished through gestures such as pointing, or, even better for focusing someone's attention, a combination of word and gesture.[64] Deixis, in other words, injects context into communication.

Grammatically and rhetorically, deictic words reinforce the chronotopic work of language by anchoring listeners in particular temporalities and places, shoring up references. Allison M. Prasch elaborates a "rhetorical theory of deixis," showing how rhetoric can help draw out the political and ideological underpinnings of deixis, and at the same time how deixis "thickens our contextual understanding of the rhetorical situation by adding bodies, physical locations, and temporalities into the equation."[65] Seemingly innocuous words—personal pronouns, demonstrative determiners, temporal indicators—do major lifting for chronotopes; they effectively move listeners around a "deictic field" by directing their attention. For Prasch, whereas rhetorical deixis inheres in those tiny demonstrative words, it also adheres to a speech's own place and time. Rhetorical deixis, that is, takes advantage of what the ecocritic Lawrence Buell characterizes as the elasticity of *thereness*.[66] And, as the linguistic anthropologist Kristina Wirtz shows, chronotopes work deictically to fill out the sensory experience of time and place.[67]

The deictic work of glaciers is exemplified in remarks Barack Obama delivered in 2017 in front of the Exit Glacier in Alaska: "This is as good of a signpost of what we are dealing with when it comes to climate change as just about anything."[68] Obama's remarks end by repeating the deictic word: "It is spectacular, though. And we want to make sure our grandkids can see this." Both uses of the demonstrative "this" are textbook examples of linguistic deixis: the word depends on the physical context for its meaning—the word "this" points to the glacier's visible recession, making use, in the first instance, of the visible markers to which Obama points, which indicate the glacier's extent in 1917 and 1951. Rhetorically, Obama's "this" is intensified by the looming glacier, the anticipation of what will be lost, what will someday not be seen.

Obama was not the first to seek the rhetorical deixis of a physical glacier. Al Gore did so in a 1997 speech he delivered in front of the Grinnell glacier at Glacier National Park in Montana. That speech began in this way: "I thank all of you for joining me here in Glacier National Park— one of the greatest glories of America's park system. The rich landscape we see all around us—the deep valleys and dramatic summits—date back more than a billion years, when Ice Age glaciers cut through this terrain, shaping and sculpting what is now one of the largest wild areas in the United States."[69]

In these two speeches, the setting provides both the occasion and the evidence. The Exit and Grinnell glaciers rise dramatically in the background, their recession markers bearing witness to accelerated warming. As with the ceremony remembering Okjökull, these glaciers do more than provide a spectacular backdrop for remarks on glaciers; in addition to the "this"—the substance or even, in Obama's case, the deictic referent—the glacier chronotope carries the rhetorical deictic force of *next* if just action is not taken now. The prospect of their disappearance conveys the magnitude of the climate crisis.

A plea composed by Iceland's prime minister, Katrín Jakobsdóttir, and circulated ahead of the ceremony on Ok encourages readers to see glaciers as "visual thermometers."[70] She worries about the future of an Iceland with no ice. But her short piece swiftly turns from Iceland to the rest of the globe, with predictions based on degrees of warming: "Even if emissions magically came to an end today, tropical glaciers—found in places such as the Andes Mountains and in East Africa—may not be saved. Mid-latitude glaciers may survive 2.7 degrees Fahrenheit warming but not 3.6 degrees. Most of the earth's nearly 200,000 glaciers will belong to history books, just like Ok, unless we do something about it and we do it now."[71] By the end of the opinion piece, Jakobsdóttir's focus on glaciers elsewhere widens the implications of Ok's loss beyond Iceland and even beyond the Arctic, to the rest of the planet, making the remote more visible and accessible.

Writing less figuratively, in more staid yet still emphatic prose, the geographer Michael Zemp and his collaborators present glaciers as "high confidence indicators of climate change."[72] Powerful images of these indicators have been shared across media forms to facilitate different kinds of witnessing. Anita Lam and Matthew Tegelberg examine the on-the-ground photographs of glacial retreat provided by the Extreme Ice Survey. The photographs, they argue, create "vicarious witnesses," which is to say that they facilitate witnessing through their "demonstrative representation."[73] Lam and Tegelberg's work is premised on a communication model that treats glaciers as visual evidence, leaving the witnessing to the humans interacting with the mediated storytelling approaches that they usefully identify. As a result, Lam and Tegelberg stop short of figuring the glaciers as themselves witnesses. And yet, as I noted in the introduction to this book, expanding the acts of witnessing to include nonhumans brings forth a potent form of elemental witnessing that refigures human relations to such materials as ice.

Elemental Witnessing

What is lost with the Okjökull glacier, and by extension (or really, to better align the elements, by ablation), the prospective loss of the rest of the planet's glaciers, is not just an outwardly visible part of a nation's landscape, but a witness, and potentially a whole mass of gleaming witnesses, to earth-changing events of the deep and near past such as the Little Ice Age, volcanic explosions, spikes in the atmosphere's radiation levels coinciding with the Chernobyl disaster, and finally, the rapidly accelerated warming at the hands of humans, nations, and lax environmental restrictions. Glaciers are, in this way, figured as nonhuman witnesses in popular articles as well as scientific research articles.[74] An obituary for Okjökull published in *The Economist* remembers the glacier precisely as witness to Iceland's history: "Though Ok was never big enough to have a proper gouging snout, it nonetheless did its small bit to carve out Iceland, a country where every feature of the landscape had history, and a tale, embedded in it. It stood as witness to that history, too."[75] As Howe puts it, "frozen within [these glaciers] are histories of the atmosphere."[76] Writing for the *New Yorker* after having interviewed the glaciologist Sigurðsson in the days leading up to the ceremony on Ok mountain, Lacy M. Johnson notes that "having 'memory' is just one of the many ways scientists refer to glaciers in terms that make them seem alive."[77]

Ascribing memory to glaciers or figuring them as witnesses does more than animate or even anthropomorphize them. It expands the act of witnessing to nonhuman or elemental entities. As Peters observes, "things, after all, can bear witness."[78] That things can bear witness matters for those scholars in rhetoric who have been working to articulate a posthuman rhetoric. And posthuman rhetoric matters for a world in which care can extend through, to, and beyond the human, as I discussed in the introduction and will detail further in chapter 5.

The glacier chronotope is constituted by elemental witnessing. The elemental constitution of glaciers makes them storehouses. Beneath their bright albedinous surfaces lie elements frozen into the layers of accumulated snow, formed annually as the year's snow compacts into ice, further compacting the previous year's snowfall, and so on. Telltale elements become trapped. Because glacial ice originated in the atmosphere, it contains a considerable amount of air.[79] Air bubbles, along with debris, sediment, and even volcanic ash, are taken into the ice from the atmosphere and from the ground as the glacier creeps along its path. These elements "bear witness," as one glaciologist puts it, to deep time/place, providing what scientists call proxy data about centuries— millennia, even—of climate change, cataclysmic events on both a local and, in the case of Chernobyl, a regional or even global scale.[80] One recent study published in the journal *Nature* used pre-industrial era ice core data to show that greenhouse gas emissions from fossil fuel extraction are far greater than originally estimated; in other words, glaciers bear witness to their own decline at the hands of humans.[81] They also bear witness to their own coming demise, showing forth as they do an accelerated rate of melt in recent decades. The magnitude and the rapidity of the mass loss "are unmatched in the time period of observational records, or even of recorded history," as one team of glaciologists noted after an extensive study of available glacial mass data worldwide.[82]

Communicating Magnitude

All of these potential losses—of elemental witnesses, of cultural history, and of visible horizons—condense into the glacier chronotope. The magnitude of that loss can be difficult to fathom. But Magnason, the Icelandic writer who, again, composed the text of Ok's memorial plaque, confronts the challenge of helping his readers comprehend that magnitude. Writing in the *Guardian* in the days leading up to the ceremony on Ok, Magnason uses the glacier chronotope to convey the broader and more profound implications of Okjökull's loss:

> Almost every glacier on the planet has stopped growing and most are shrinking at an alarming rate. Ok Glacier is the first in Iceland to be formally declared dead ice. In the Himalayas, Greenland, the Alps and Iceland, the glaciers are all melting. . . .
>
> According to current trends, all glaciers in Iceland will disappear in the next 200 years. So the plaque for Ok could be the first of 400 in Iceland alone. The glacier Snæfellsjökull, where Jules Verne began his Journey to the Centre of the Earth, is likely to be gone in the next 30 years and that will be a significant loss. This glacier is for Iceland what Fuji is for Japan.[83]

Magnason effectively spreads a chronotopic network around the planet, as if spinning a raised relief globe and skimming his palm over its ranges and peaks. He moves from the Alps, which have about 4,000 glaciers, all of which are losing ice, to Mount Fuji, a heritage site with no glaciers that is nevertheless a visual icon for its nation.[84] Imagine, Magnason implicitly asks readers, if Japan were to lose Mount Fuji. In conjuring and connecting regional imaginings, Magnason creates what Kendall Phillips and G. Mitchell Reyes might call a global memoryscape, an imagined world that, in this case, begins in a region but extends to far-flung distances.[85] Only instead of a memoryscape, this one is something of a futurescape, one that seeks connection through the subjunctive sentiment of the "what if," visually conjuring imaginative relations by drawing imaginary lines from peak to distant peak.

Magnason frequently aims to reduce distance by making explicit connections. In the same *Guardian* piece, he uses the familiar (at least to cosmopolitan readers) image of the Empire State Building to convey the magnitude of "thickest packs" of glaciers; in the case of mighty Vatnajökull in eastern Iceland, these packs run 1,000 meters deep. To convey the magnitude of that density, Magnason asks his readers to "imagine stacking three Empire State Buildings on top of each other—then stretch that bulk over the horizon." He adds, "To think that something so huge is actually fragile is beyond comprehension."[86] In Howe and Boyer's documentary *Not Ok (A Little Movie about a Small Glacier at the End of the World)*, Magnason tries a similar move. This time, rather than invoking a recognizable and famously tall architectural structure, he works with the size of Iceland itself, over which he tamps and spreads Iceland's glacial mass: "Glaciers are of course these mighty structures on our landscape. And actually if Iceland was covered with glacier, like if it was evenly spread over the island, Iceland would be under 20 meters of ice. Yeah. So actually a very significant part of the mass of the island is ice and snow. But because it's packed in these mountains we don't maybe feel them in everyday life."[87] Magnason tries to help people comprehend the magnitude—in this case, the mass and weight—of glaciers by having them feel it, by bringing the accumulated ice and snow to where they live and move.

Whether connecting famous peaks around the globe, stacking or multiplying and stringing together a familiar architectural object, or flattening Iceland's great and looming icecaps to bring the ice and snow under people's feet, Magnason is tapping into imaginative faculties by translating enormous ice packs into units knowable and familiar. In doing so, he tries to make these structures *felt* in all their might and weight. His scalar sliding seeks to make the distant proximate and the unfathomable more fathomable, to convey the magnitude, the significance, the weight of the melting glaciers. Magnason makes the magnitude of glacial loss more accessible for Icelanders and non-Icelanders alike, an effort that he replicates in a generational chronotope out of which he fashions what he calls "intimate time."

Intimate Time

Magnason has made it his mission to help people better imagine long futures. Lately, when he serves as a guest lecturer, say, or appears on podcasts, or writes an opinion piece, he asks his audience members to join him in a version of the following exercise: "Let's look at the future. Let's not think of 3-D goggles or this gimmick stuff. Not even A.I. Just look at the future, and just imagine, when is somebody still alive that you will love?" Magnason continues, with almost hypnotic repetition of the second person "you": "You will become ninety. Your favorite twenty-year-old in your life will be born in 2070. It will be a person you have held as an infant. You have raised or have participated in creating this person, this character. And so the person that will be closest to you in 2090 is born in 2070. And when is that person still talking about you as their main person in life?"[88] In this account of the exercise, Magnason goes on to describe how another version of this thought experiment went just prior to the interview when he was addressing a group of twenty-year-olds. As he describes it, the young adults "do the calculation. Okay, 2070, the person is ninety in the year 2160." Then Magnason tells them, "Okay, this is your intimate time. This is the time that belongs to you." Magnason ends the exercise by layering that intimate time with the specific timeframe of the latest climate predictions: "Then

I ask them 'Okay, now, how about 2070? What do you think about this [IPCC] report on 2100? Do you feel it's beyond your imagination?'" Magnason's exercise helps people stretch their imaginations beyond the now or the near future, beyond even the span of a single life, where so many imaginative capacities stall. The phrase "not in my lifetime" too often terminates the available imagining, full stop. Magnason's exercise presses past the lifetime through a generational chronotope made palpable in imagined physical contact: "It will be a person you have held as an infant." That is, he takes on what might be the most difficult rhetorical challenge of climate change, even more difficult than reducing the unimaginable distances to Earth's most remote glaciers: the challenge of timescale.

And Magnason's intimate time, dependent as it is on a generational chronotope that stresses the *chronos* part of the word, traverses a vast expanse of time with an imagined chain of lifespans, a chain whose links are felt as what he calls love, strengthened by imagined touch. As he puts it in a written account of the exercise, "That is a time whose meaning they can touch with their bare hands."[89] He effectively addresses a problem of imagination by using an age-old imaginative tool: the thought experiment. As Nathan Crick has shown, the form of the thought experiment allows speakers and audiences to create "an entirely *new* picture between them," and as an argumentative form can allow a wide range of participation, inviting in as it does individual lived realities.[90]

Magnason is not, of course, the first to believe that the problem of climate change is an imaginative one. In fact, as I discussed in this book's introduction, environmental scholars have long pinpointed limited imaginative capacities as a major obstacle to addressing the climate crisis. Magnason's intimate time enacts what Lawrence Buell, for example, calls "environmental connectedness," and it openly struggles against what Buell calls "the limits of habitually foreshortened environmental perception."[91] Nicotra and Parrish, working with the idea of chronotopes, label this foreshortening of perception "time-space compression," a kind of time that governs and promulgates neoliberal ideology, and they offer deep time of the past (paleoclimatology) as a way to offset such a myopic view of time. As Nicotra and Parrish put it, "Apart from the few (like paleoclimatologists) who have become accustomed to working with vast scales of time as part of their regular work, the deep time perspective typically does not come naturally for those of us who inhabit the 24/7 world that characterizes so much of contemporary Western life."[92] Magnason tries to make vast expanses of time more accessible not through disciplinary training, but through intimate, familial ties, through imagined feelings of love and of touch.

Magnason tries, repeatedly, to stretch imaginations further and further futureward. His generational enchainment accomplishes two important rhetorical feats. First, rather than depending on the past to condition imagination (though sometimes the exercise begins with his recounting his relationship with his own grandmother), he opts to stretch imagination's tendrils far into the future, and he does so in a way that acknowledges and carries feeling—namely, familial (and perhaps familiar) feelings of love and care. Magnason, that is, performs the hope articulated by Buell and Pezzullo: the hope for wrapping attention with care. In doing so, he activates what Nicotra and Parrish call time's "affective dimension."[93] Magnason's intimate time therefore vivifies the inclusive spirit of Bakhtin's chronotope. Rather than "jostling" with each other, as Nicotra and Parrish put it, these chronotopes coexist and mutually reinforce each other. The glacier chronotope can persist right alongside, and indeed link directly to, the generational one.

Magnason's exercise may also prompt those of us studying rhetorical theory to rethink our presumptions about rhetoric's time. For about two decades, theorists of rhetoric (myself included) have favored the version of time the Greeks labeled *kairos*. *Kairos*, as students of rhetoric usually learn early on, names the quality or rightness of time, time as opportunity. *Kairos*'s chronotope, if it can be said to have one, is the here and now, or in maxim terms, the now or never, the knife's edge of decision. The appeal of *kairos* is evident; I often find it to be the Greek term that holds the most salience for college students learning about rhetoric. Occasion and opportunity (the meanings stressed by *kairos*'s Latin equivalent, *opportunitas*) are exciting; they are palpable, they press on a moment and call for a response. Even for those of us who write rhetoric's histories, *kairos* is vastly more appealing than the kind of time that is customarily presented as its counterpart: *chronos*, the plodding and incremental time of clocks and calendars.

These two versions of time divide the quality (*kairos*) and quantity (*chronos*) of labor, emphasis on divide. And yet in privileging the opportune moment, the buzzing draw of immanence, rhetoric scholars have perhaps made *chronos* into a vast expanse, treated it as an unexamined, unimportant, neutral backdrop. *Kairos* is still, of course, the time of contingency, rhetoric's time.[94] And *kairos*'s urgency is still crucial for taking action now. But as Magnason's exercise shows, particularly in the context of the climate crisis, the "or never" part of the maxim demands an ability to engage *chronos* too—to see the long view, what is (or is not, or might well not be) next.

Here, then, is one important way that the climate crisis is changing rhetoric: as an issue that draws on data from the deep time of epochs and -cenes, whose implications stretch far futureward, the rapidly changing climate exposes the problems with leaning too much toward the now and the immediate, and asks instead that we find ways to stretch each other's imaginations so that they can hold open multiple time-scales simultaneously, especially the now of action as it relates to the long future of those actions' implications. Magnason's remarks on Ok included an adapted version of his time exercise, which Howe framed to me as "creating generational connection":

> Andri said some words too about you know the generations, and one of the things he's very interested in is this idea of someone you know, not a baby, but like someone you get to know so say like your grand-child, you get to know them, you love them, how in their timeframe like how old they get to be as old as your grandparents were you know what year will it be when they are also spending time with someone young that they love. So he said something at the top of the moun-tain like there's a young person here today that will be speaking to her grandchild one day and will tell her grandchild what we did here on top of this mountain, and that grandchild will go on to tell her grand-child maybe, kind of creating that generational connection.[95]

Here Magnason's reference to the young people present on Ok that day draws deictic power from the youth who had climbed the mountain, carrying protest signs (one sign, visible in the press photographs, reads "Declare Emergency Now"). In a somewhat familiar move, the chil-dren at the ceremony became figures of the future, the most proximate link, by way of Magnason's generational chains, to the otherwise un-imaginable date of 2100. What's important here is that words help feel-ings form the chains: they work, that is, to stimulate a vision of future tellings, more words. Words themselves stretch across time and space, pulling imaginations with them, much like the Empire State Build-ing that Magnason replicated and stacked to convey the magnitude of the masses of ice beneath Vatnajökull. These imagined retellings hold his chronotope of intimate, generational time alongside the palpable now—the lapsed glacier. The living children join, as themselves wit-nesses now and in the future, to urge a somewhat paradoxical vision of future memory.

Magnason's version of intimate time stitches together generational units of time with familial love, thereby extending a lifespan, usually a limiting factor for individual caring, past death, with blood and DNA acting as conduit. Howe's version of intimate time makes available a bond that is more communal, one less dependent on genetics or biological kinship *per se*: the ancestry model of the future made by people indigenous to North America. In our conversation, Howe described making such a point in her remarks at the summit: "I wanted to remind people of something that we know from Native American philosophers and thinkers was this idea of being better ancestors and being a good ancestor, so I mentioned that as an important way forward. If we could be better ancestors, then we could imagine a better future." Here, Howe takes a familiar chronotope—that of ancestry, typically a category of history, often lined up right next to something like "forefathers" or "founders" (in American discourse), and flips that chronotope to face the future. Her planned remarks use this phrasing: "As many indigenous friends and colleagues have put it, 'we need to learn to be better ancestors.' Becoming a better ancestor for future generations, means acting now."[96] The result is this: those living now, and specifically those on Ok mountain that day, could start to see themselves as future ancestors, humans whose actions now matter with greater intensity later. In this way, the chronotopes created and shared in the Ok ceremony allow the participants to imaginatively inhabit multiple timescales simultaneously, and to do so with intimacy and care.

Memorial Witnessing

As with the ceremony, place matters for the memorial too. The surrounding glaciers exert the deictic pull discussed above; they serve as elemental witnesses to the ever shifting now. But in the case of the memorial, another elemental witness deserves attention for its centrality to both the ceremony and the memorial, for its almost profound durability, for its own timescale—far longer than lives of individual humans, longer than the projected remaining time of the dwindling glaciers, and yet probably not as long as the plaque itself: the large stone on which Ok's memorial plaque is now mounted (figure 2.4). Howe and Boyer, along with the people who accompanied them on the "Unglacier Tour" the year before, following the premier of their documentary film *Not OK*, took as part of their mission to identify an ideal location for the memorial plaque. Here is how Howe and Boyer put it in the proposal:

As part of the Unglacier Tour we asked the participants to help us locate a suitable site for a potential memorial plaque. There was unanimous agreement among the participants that the best site was a large pyramidal stone at the coordinates: N 64°35.498' W 020°52.253' and an elevation of 1.114 meters. The stone, as can be seen in the images at the end of the proposal [see figure 2.4], has several advantages: it

2.4 Stone for memorial plaque, August 2018. Courtesy of Cymene Howe.

is situated on the edge of the summit on relatively flat ground, it is already angled in such a way as to provide easy viewing, it is large enough (over 1 meter in height) to mount a sizable plaque on, and it has a natural rock formation near it that resembles a bench where hikers could comfortably pause to contemplate the memorial.

When Howe described to me how they found the rock on that hike, she emphasized the almost joyful certainty the group members shared when they found it: "As we were descending off the top of the summit, here is this, as someone put it, 'straight from central casting' stone that's sitting there, it's just at the right angle, it's got almost a completely flat face which you'd need to mount the marker, and it's big, and it has these other stones around it that are almost like benches or places where people can rest and meditate. So it was kind of incredible, and everyone was like 'Yes, that's the rock. That's the rock. That's the rock.'"[97]

That rock, now plotted, measured, and drilled, bears the bronze plaque on which Magnason's letter to the future is etched. The fastening of the plaque to a rock on Ok draws into the ceremony and the memorial another inhuman element. The inhuman rock and the human participants become "partners in feeling," to return to a phrase which in a previous book I derived from Aristotle.[98] Jeffrey Jerome Cohen's book *Stone* presents rocks similarly, as inhuman partners with which "we build epistemological structures that may topple upon us."[99] Stones have been, as Cohen's extended meditation shows, pressed into our very epistemes as tropes of silence, strength, cold. And yet stone is, in his words, a "communication device that carries into distant futures the archive of a past otherwise lost."[100] Rocks work in their own way, of course, as elemental witnesses, bearing fossilized imprints left by species long extinct. Cohen's reflections on "lithic agency" require attention to a time that is slower and deeper than the timescales and speeds that humans are conditioned to comprehend. He therefore discerns "in stone not just a rebuke to our anthropocentrism" but a liveliness, what he calls a "vivacity," that may have things to say.[101]

The rock the group selected is basalt, which, like the rocks all around it, including the ones the ceremony participants picked their way through on the way up the mountain, is relatively porous, essentially cooled lava. According to Howe, the metalsmith who fashioned the plaque for the memorial "said that the plaque would long long long outlive the rock onto which it is placed." Howe then shared with me her own vision that "in 1,000 years or something, the plaque will just be sitting on the ground; the rock will have disintegrated."[102] Much like glaciers, rocks bear inhuman temporalities that need to be heeded if the "slow-motion collapse," to use Boyer's phrase, is to be further slowed or even averted. Layered onto this elemental chronotope is a small collection of words, etched in bronze, that counter the stony silence: a letter to the future. Basalt bears bronze, which in turn bears words, words that invoke a future public memory.

Here are the words etched into bronze:

A letter to the future
Ok is the first Icelandic glacier to lose its status as a glacier.
In the next 200 years all our glaciers are expected to follow the same path.
This monument is to acknowledge that we know
what is happening and what needs to be done.
Only you know if we did it.
Ágúst 2019
415ppm CO_2[103]

The plaque's words hardly need commentary—most of the lines here work in epideictic fashion to document the loss of Okjökull and to witness what that loss portends. The glacial chronotope guides the first three lines, joining with a chronotope of the path in the second line. The path, here, somewhat like the chronotope of the road of which Bakhtin is so fond, is that which "passes through *familiar territory*."[104] Paths, after all, must be forged, if not worn, and Ok has led the way. The monument, though, uses the memory of Ok and the chronotope of the path to amplify and bear witness to the fact of awareness. This monument, in other words, memorializes Ok, but it also memorializes the self-conscious Anthropocene, magnifies its importance, stands as a stony witness to what humans know now.

The memorial to Ok records that knowledge by adding as documentary evidence the known level of carbon in the Earth's atmosphere, which soared to an unprecedented 415 parts per million three months prior to the ceremony, prompting one meteorologist and climate journalist to proclaim, "We don't know a planet like this."[105] We don't know such a planet, and yet as the planet and the plaque on Ok endure, we decidedly will have known it. The unprecedented, that is, is and will continue to be an ever renewing condition unless humans act on the other knowledge documented in the memorial: that of "what needs to be done."

The "what needs to be done" launches the letter's last line, which expands from the documentary work linking Ok to the world's remaining glaciers through the chronotope of the path, and invokes a different chronotope altogether: that of the reader, specifically, a future reader or readers.[106] The last line of the letter—"only you know if we did it"—slips instantly into an as yet unknown future. The first-person plural "we" remains constant from the previous two lines ("we know what is happening and what needs to be done"), but the pronoun "you" appears abruptly, and with it, a scene of future judgment. As Cymene Howe put it in an interview with Peter O'Dowd, "The second person plural form—the you form—is a very direct way of speaking to people. When reading the plaque you can . . . imagine this future person standing there reading over this memorial, and so, by addressing this letter to the future, by invoking the "you," we're trying to very explicitly draw attention to that potential and that future."[107] The moment of the "you" works, as Howe put it to me, "all of a sudden." The chronotope of the reader conjures a vision of the future, snapping an audience into view, in front of the bronze plaque, even perhaps faintly reflected in its burnished surface. By means of the chronotope of the future reader, to use Bakhtin's words, "time thickens, takes on flesh, becomes artistically visible."[108] The pronoun/perspective shift and the shift of verb tense (from the present "we know" to the conditional past "if we did it") carries the "charge" of time/place: it tries to jolt now-readers into expanding their scalar imaginations. The "you," that is, entwines *kairos* with *chronos*, and it does so in a flash. By conjuring future readers of what effectively amounts to a date-stamped admission that humans are knowingly destroying the planet, the last line introduces a vision of future public memory of what is happening now. Witnesses, elemental, documentary, and imagined, layer onto each other like the compacted snow of glaciers themselves. And herein lies the rhetorical possibility of chronotopes when turned toward the future in the context of epideictic rhetoric: timescales multiply; possible futures zoom into view, and with them the prospects of further loss, judgment, and blame.[109] Care can infuse attention in this way, holding open the possibility of different paths. We know, after all, what needs to be done.

Conclusion: Ceremonial Magnitude

A set of ritual practices, a pointed letter engraved in bronze. The distinct materialities of the Ok ceremony and its memorial plaque reflect the equally distinct bodies of literature in public memory studies in rhetoric—those that on the one hand lean toward public address, the subfield within which scholars examine and discuss ceremony and modes of address, and on the other hand incline to memorial studies, the category for studies of material monuments.[110] Memorial services and memorials are, of course, nothing new. But their use for a fallen glacier is, as Magnason noted in his framing remarks for the Ok ceremony, without precedent. Howe and Boyer stressed this novelty in their proposal to the municipality: "We would like to emphasize that this memorial plaque would actually be the first of its kind in the world. Nowhere else has a fallen glacier been memorialized. But with so many glaciers now facing existential threats, we think the Ok memorial could serve as a model for others to come."[111] The choice Howe and Boyer made to hold a ceremony and install a plaque gave new purpose to familiar rituals, genres, artifacts. Together, the ceremony and the memorial work in the documentary realm of epideictic rhetoric to shape deliberative and political possibility by helping people imagine futures on multiple timescales at once.

Together the ceremony and memorial exert a pressure that differs from that exerted by, say, a protest, though the presence of a few protest signs at the ceremony added elements of demonstration. The ceremony and memorial offer, to borrow a phrase from a brilliant essay on magnitude by the rhetoric scholar Stephanie Larson, means of "interrupting apathy."[112] And they do so not by overwhelming people with scientific knowledge, but by presuming that knowledge, and by recording the fact of humans' awareness. The plaque bears spare reminders of the unprecedented nature of things: 415ppm CO_2, Ok's loss. And the chronotopes offered by the ceremony and the memorial help people to take in the magnitude of the climate crisis bit by bit. As I have shown, the chronotopes of wishes, glaciers, generations, paths, and readers amplify and bring close what is, for Howe and Boyer, a story about a "small glacier at the end of the world."[113] Their action is wish-bound and attentional, and as I have stressed, the attention's quality, its care, and its temporal flexibility make all the difference.

At one particularly potent moment in the ceremony on Ok, the ceremonial and the material memorial came together. That moment is beautifully recounted by the Houston-based writer Johnson, who attended the ceremony: "Howe and Boyer asked the children to come to the front of the crowd. 'We need to understand our relationship to the world in ways we haven't had to in the past,' Howe said. 'We need to be able to imagine a new future.' There was a moment of silence as the children pushed the plaque into place. The day had cleared a little, and I could see across the Kaldidalur to the glaciers on the opposite peaks. Below them, in the valley's deepest crevice, a meltwater lake was forming, already so blue and deep."[114] Johnson's freeze-frame ekphrasis of this quiet moment of clearing conveys the stunning deictic work of the visible glaciers in the distance together with the children, the next of Magnason's linked lifespans. And Howe's words—"we need to be able to imagine a new future"—aptly characterize the rhetorical, subjunctive, wishful labor of both the ceremony and the plaque, which model even as they perform the imaginative labor before us all.

Of course, youth do more than help imagine a future. More and more, they are themselves stepping in to remind leaders and citizens that they—the young generations—must live the future looming before them, and in doing so, they are developing a new future-oriented rhetoric, one at once grounded in scientific knowledge and charged with feeling. Chapter 3 will turn to the rhetorical practices of a band of youth rhetors who are, much like the creators of the Ok ceremony and memorial, venturing into unprecedented rhetorical terrain.

CHAPTER 3

"In a World Full of 'Ifs'"

THE FELT TIME OF YOUTH
CLIMATE RHETORS

On September 18, 2019, as the chairman of the Foreign Affairs committee in the US House of Representatives welcomed everyone to a hearing on the climate crisis, Room 2172 of the Rayburn House Office Building was thrumming with energy. As the hearing got underway, the dais still filling with members of the subcommittee on Europe, Eurasia, Energy, and the Environment and members of the House Select Committee on the Climate Crisis, about twenty-odd camera people knelt in front of the witness table, their backs to the legislators, pointing their cameras at the first witness, the Swedish climate activist Greta Thunberg, who was taking her seat (see figure 3.1).

As youth climate activists go, Thunberg's fame is unmatched. Scholars, journalists, and filmmakers love her brazen, "unvarnished" delivery.[1] As well they should. *Time* named her "person of the year" in 2019.[2] Just a day prior to the hearing, Barack Obama called her "one of our planet's greatest advocates."[3] The congressional hearing took place two days before the "Global Week for Future," a set of worldwide strikes and protests led by "Fridays for Future," the school strike for climate effort initiated by Thunberg. During her eleven-week stay in North America, Thunberg made guest appearances on *The Daily Show with Trevor Noah* in New York and *The Ellen DeGeneres Show* in California; she led rallies and delivered speeches, including an electric four-and-a-half-minute speech, a stunning blend of fact and feeling, to world leaders at the UN's Climate Action Summit four days after this hearing.[4] But at this congressional hearing, Thunberg stripped her remarks down to bare science. Her congressional testimony consisted of her entering into the record the 2018 IPCC Special Report on Global Warming of 1.5 degrees Celsius and leaning into the microphone to speak these words: "I don't want you to listen to me, I want you to listen to the scientists."[5] Her

3.1 All (camera) eyes on Thunberg. Still frame from the beginning of the September 18, 2019, hearing, House Foreign Affairs Committee.

opening statement, seven short sentences, lasted 51 seconds. The idea of a sixteen-year-old girl using her voice to amplify climate science is, of course, as newsworthy as it is laudable. Given Thunberg's renown, and given her head-turning boldness, it is no surprise that the vast majority of the news headlines focused, like the cameras, on her.[6]

But rhetorically, Thunberg's science-only testimony, a perfect instance of what Timothy Morton has termed an "information dump," landed with a thud, foreclosing deliberation and sealing itself within scientific data.[7] Jamie Margolin, the witness sitting to Thunberg's left that day, seems to have picked up on the reception of Thunberg's report. She told the *New York Times*, "She gives very blunt observations. If I had done that, people would've thought I was an entitled brat."[8] It is customary to include witnesses' written testimony in the hearing transcript, and so the absence from the hearing transcript of the IPCC report that Thunberg entered into the record is noteworthy. Elsewhere, Thunberg delivers a fiery generational rhetoric of shame and panic, one that, in truth, I initially planned to investigate in this chapter.[9] Yet when placed into the context of youth climate witnesses at congressional hearings, Thunberg's remarks at this particular hearing recede into the background. The IPCC report and other gold-standard climate science like it served as the first premise for most representatives on the dais.

Such laser focus on Thunberg to the exclusion of the other youth witnesses therefore misses a potent rhetoric put forward by Thunberg's American climate activist counterparts, including those accompanying her at the witness table that day, and witnesses who appeared before the Select Committee on the Climate Crisis earlier that same year, on April 4.[10] The other witnesses at these two hearings present a powerful rhetoric, a necessary rhetoric, one that Thunberg, in this instance at least, refused.[11] As a group, youth climate leaders have a good deal to contribute to this book's question about how climate change is changing rhetoric, mainly because they are often the ones enacting that change. Young people are at the global climate justice movement's forefront: walking out of class, leading rallies, mobilizing through social media. In addition to organizing demonstrations, they give TED talks, they bring lawsuits, and they testify in front of deliberative, policy-making bodies (e.g., British Parliament, the United Nations, and, as in these cases, the US Congress).[12] And they are making a difference, too. Emmanuel Macron, the president of France, said in an interview about young climate activists, "They helped me change."[13]

Chapter 2 elaborated how chronotopes intensified the quality of imaginative attention in the epideictic context of a memorial ceremony and plaque. This chapter moves to the deliberative, policy-making context of congressional hearings and finds there a set of felt temporal rhetorics circulated by young climate rhetors. These temporal rhetorics, a potent mix of fact and feeling, came to the fore at the inaugural hearing of the House Select Committee on the Climate Crisis. This new committee was authorized on January 9, 2019, by House Resolution 6, and its charge was "to investigate, study, make findings, and develop recommendations on policies, strategies, and innovations to achieve substantial and permanent reductions in pollution and other activities that contribute to the climate crisis which will honor our responsibility to be good stewards of the planet for future generations."[14] The Select Committee's first hearing took its cue from the clause in its charge about stewardship and future generations.[15] That hearing was titled "Generation Climate: Young Leaders Urge Climate Action Now," and its witnesses were Chris Suggs, from Kinston, North Carolina; Lindsay Cooper, from New Orleans, Louisiana; Aji Piper, from Seattle, Washington; and Melody Zhang, from Detroit, Michigan.

The second hearing featuring youth witnesses on the matter of climate was the one with Thunberg. Chaired by Congressman Bill Keating, the hearing was held jointly by Keating's subcommittee and the Select Committee. Thunberg, who had arrived in the US three weeks earlier on a no-emission sailboat, was accompanied by Jamie Margolin, from Seattle, Washington; Vic Barrett, from White Plains, New York; and Benji Backer, also from Seattle.

In order to provide a better sense of the climate rhetoric put forward by the younger generation, this chapter will parse the rhetorical approaches of four youth witnesses at these two hearings. I chose to focus on these two hearings rather than on social media posts, rally speeches, editorials, or books written by youth rhetors, first, because in the halls of Congress, youth rhetors are explicitly addressed by legislators as witnesses for the future. After bringing the September 18 hearing to order, the subcommittee chair, Congressman Bill Keating, told his colleagues, "we are speaking to the future."[16] As witnesses, these particular youth enact the conception of witnessing I laid out in this book's introduction—weighty assertions of material presence that lay bare injustices and demand a reckoning. They focus unflinchingly on morality and justice; they stretch witnessing's temporality toward the future; they hold people to account and call for a reckoning. They assert an embodied, lived authority on how their future feels.

Second, the youths' congressional testimony is of particular interest because with it, they directly address lawmakers, those whose actions they need to "urge," to borrow the verb from the first hearing title— and fast. In his *History of the House of Representatives*, the congressional historian George Galloway calls the committee rooms "the meeting ground between the public and the Congress."[17] The rhetoric scholar Daniel Brouwer characterizes the public constituted by hearings as a "weak public nested within the strong procedural public of Congress."[18] The two hearings I examine show a rising future-facing public comprising young citizens, and as such they offer glimpses of how that emergent public is framed and received by committee chairs and members.

Of the eight witnesses who testified at these two hearings, I focus on four in particular—Suggs, Piper, Barrett, and Margolin—because these four bring to the witness table identities and experiences that carry them closer to the crisis than the members of Congress they are addressing. Suggs and Piper are Black, Barrett is Black and Latinx, and Margolin identifies as queer and (white) Latina. As Barrett put it in an op-ed published by the *Guardian* two days after the hearing, "the identities I inhabit: first-generation, trans, indigenous, Latinx, black, youth; all make me uniquely vulnerable to the impacts of the climate crisis."[19] Barrett's inclusion of "youth" in the list of his multiple identities lays the groundwork for an intersectional climate justice that buttresses a rhetoric of felt time, a new rhetorical mood. Together, these witnesses present the climate crisis as a tangle of profound injustices wound with intense temporal feeling. These youth, that is, bear witness to the future, and they do so by describing how it feels to face a future so foreboding.

Witnesses play a complex role in congressional decision making, one that a rhetorical perspective is well positioned to draw out. As Brouwer puts it, witnesses can "provide members with rhetorical resources during their discursive performances on the floors," thereby bolstering a lawmaker's case for a particular policy approach.[20] In these two climate hearings, though, what the majority members needed was help convincing intransigent, partisan colleagues to agree to take bold action, and that rhetorical feat would require more than a scientific report. As Select Committee Chair Congresswoman Kathy Castor put in the transition to the spring hearing's question-and-answer session, "I really appreciate everything that you have said today to help spur us on to action and kick off our first hearing."[21] The youth witnesses, that is, were invited to both hearings to set a tone, in what seems like a last-resort effort to move their colleagues. On this point, Congressman Bill Keating's opening statement at the second hearing was explicit: "I hope your testimony today will galvanize us to act now before we are truly out of time."[22] Congressman Keating and Congresswoman Castor both wanted the young witnesses to help them help their colleagues feel the urgency of the climate crisis, to jar them out of their partisan grooves. The verb Keating chose, "galvanize," is, of course, a term from electricity, derived from a chemical current called a "galvanic current," a rousing, jolting force.[23]

That jarring, pulsing charge aptly characterizes the rhetoric of American youth climate leaders whose testimony this chapter examines. The temporality they present carries with it a variety of felt experiences, ranging from alarm and despair to exhaustion, sadness, and betrayal. And that felt time contrasts sharply with Thunberg's "listen to the science" testimony at the second hearing.[24]

To elucidate the felt time of youth climate rhetors, this chapter begins by examining rhetorical scholarship on time and temporality to show how time, wound with feeling, has been an important yet latent factor in the field for some years, until very recently, when scholars have brought feeling to the surface. Then I turn to the hearing transcripts, statements, and exchanges in the hearings' question-and-answer periods to show, first, how the chairs of the committee encourage the young people to share their future-directed feeling as their mode of witness, and second, how these youth climate rhetors oblige, conveying how it feels to endure the slowly unfolding climate crisis, to contemplate the grim consequences of governmental inaction, and to have been born into a crisis whose magnitude is as yet unprecedented. What

emerges from the testimonies is a palpable sense of how the future—
this future—feels: a generational rhetoric of felt time.

The stakes of felt time are big. First, policy-making deliberation is
almost always presented as information gathering and weighing, but in
these testimonies, what is being gathered is not so much information
as it is stories, feelings, and harms.[25] For these witnesses, the stakes are
a matter of survival—survival of communities and peoples, and of an
ability to plan and imagine any kind of future. In the context of rhetoric, a theory of felt time augments prevailing theories of time in rhetoric, showing why the beloved distinction between *kairos* and *chronos*—
qualitative and quantitative time, the moment versus a minute—while
important, only scratches the surface of temporal rhetorics. Perhaps
most importantly, it is no coincidence that felt time is beginning to surface in rhetoric in matters of racial justice at around the same time that
it is in the context of the climate crisis, for the climate crisis, as is well
known, disproportionately affects Black, Indigenous, and People of
Color (BIPOC) communities.

By zeroing in on the congressional testimonies offered by BIPOC
and LGBTQ youth in particular, then, this chapter shows how the overwhelming temporality of the climate crisis feels to them awfully similar to structural racism and discrimination, and that's because the two
reinforce each other in ways that at least two of the witnesses explicitly detail. In other words, for a subset of "generation climate," being
born into a crisis not of their own making is not all that new. If, as Anna
Lowenhaupt Tsing puts it, "precarity is the condition of being vulnerable to others," then these youth are multiply precarious.[26] In exerting
a pressure unlike other pressures, climate change amplifies and draws
out those multiple precarities, and those precarities are made salient
through a generational rhetoric of felt time. Youth rhetors, particularly
the BIPOC and LGBTQ witnesses at these hearings, speak those temporal feelings freely, with clarity, and even at times with fire ignited by
a grinding friction.

Feeling Time

In 1974, the rhetoric scholar Bruce Gronbeck, toward the beginning
of an otherwise straightforward account of rhetorical timing—one of
the field's first—offers a tantalizing hedge: "To be sure, making a public
message 'timely' often is a matter of feeling and hence perhaps unan-
alyzability."[27] His qualification is itself an untimely observation, inas-
much as rhetoric scholars at the time were not yet equipped to consider
the connections among time, timing, and feeling. Now, though, rheto-

ric and feeling constitute a rising area of study at once urgent and, *contra* Gronbeck, analyzable.[28] Gronbeck's instinct here—that the art of timeliness may well be "a matter of feeling"—is, as my account of youth climate testimony will show, largely correct, even if his conjecture that feeling is somehow not subject to analysis may not be quite right. It is possible, that is, to consider how felt time becomes manifest, how it works rhetorically.

There is no time like the present, as the chestnut goes. Scholarship focusing on feeling and rhetoric is on the rise, as is scholarship focusing on temporality and rhetoric. The two in fact are beginning to merge, and these youth climate rhetors effectively complete the merger. Within the span of half a year, two journals published special issues on temporality, or on what Matthew Houdek and Kendall Phillips, in the title of the introduction to one of those special issues, call rhetoric's "temporal turn."[29] Even so, as Gronbeck's publication date suggests, and as John Poulakos's account of sophistic rhetoric noted as far back as 1984, rhetoric's concern with temporality is not new. For Poulakos, concerns of "the moment" have long been what draws forth rhetorical action in the first place: "for the most part, what compels a rhetor to speak is a sense of urgency."[30] And the word "urgency" here and elsewhere—including this book's title—comes attached to the word "sense." "A sense of urgency" carries a felt quality that rhetoric can help spread. Urgency, too, as a felt sense, sparks rhetoric.

What *is* new, then, is work that examines exactly that which Gronbeck thought might be unanalyzable: the feelings that urgency conjures, that serve to constitute, reinforce, and intensify it. Rhetoric scholars, that is, are beginning to lay bare what it means to talk about qualitative time, a feature that has long been noted as that which distinguishes *kairos*, "rhetoric's time," from *chronos*, everyone else's.[31] And feelings carried by time and urgency are coming to the fore in rhetoric scholarship focused on injustice and racism.

Urgency as felt sense is given new salience in a coda written by the contributors to one of the aforementioned special issues on temporality. The coda, composed and published in the midst of the 2020 COVID-19 pandemic, and during a summer of protests of police brutality against people of color in the United States, offers the following insight: "Temporality might be understood in this regard as existing in a state of tension with levels of privilege on the one side and awareness of precarity on the other. Those residing in their privilege may feel the temporality of the present moment as a long uncertain pause; those experiencing the approach of precarity feel time as more urgent."[32] To put it in a way that links this crucial intervention with prior work on temporality, time

feels different in a crisis, and privilege and precarity can modulate vary-
ing senses of urgency.[33]

As Tamika L. Carey points out in her brilliant article on Black wom-
en's rhetorical impatience, itself an articulation of felt time that Carey
calls a performance, conceptions of time and timing entail power structures.[34] Carey's analysis of Black women's public commentary on time
foregrounds what she calls "temporal hegemony," a situation where
"ideological and material structures converge into a culture of hostility
that pushes equity for a group further out of reach."[35] Temporal hegemony has broad utility as a concept, but its context—Black women's
rhetorical impatience—must be acknowledged and honored. Carey's
conception of rhetorical impatience captures how time can be felt—
how it is felt, and how the Black women whose performances Carey
documents refuse to abide by the hegemonic demands of patience.
That refusal, which Carey names "performances of frustration," simultaneously marks hegemonic time and reasserts a different kind of
time—time that doesn't just operate as a contextual backdrop, but is
brought front and center.[36] Indeed, Carey's piece explicitly exposes the
inadequacy and seeming impermeability of the *kairos-chronos* dyad as
the lead concepts of time in rhetoric, concepts that scholars too often
presume to be neutral or passive, even as they write of their rhetorical
potency: time shifts, exigence rises and falls.

That habitual reliance on the distinction between *chronos* and
kairos—time as it moves on, quantitative time; and time as it forms into
an occasion, a ripe, urgent moment, qualitative time—in contrast, has
largely left feeling to the side. But in 2020, the year when people reached
a breaking point on the matter of racist police brutality in the middle
of a global pandemic, the differential feeling of time reasserted itself
with an undeniable power, revealing *kairos* and *chronos* to be wound
together in ways that have for the most part gone unacknowledged in
the scholarship on rhetoric and temporality.[37] Temporality, as Carey
and the authors of the coda suggest, is by no means unfelt, especially
by marginalized peoples. As Carey puts it, "a person who understands
the reality of intersecting and perpetuating oppressions also knows that
there is almost always a viable rhetorical situation requiring action."[38]
The perpetual sense of urgency is what draws forth the feeling of impatience that Carey finds so forcefully performed by contemporary Black
women.

The rhetorical impatience Carey delineates might be aptly characterized as a "rhetorical mood." This phrase appears in rhetoric and cultural
scholarship more or less fleetingly. Lauren Berlant uses it in an early essay "Trauma and Ineloquence."[39] "Rhetorical mood" hovers and looms,

both taking on and spreading feeling and affect. The phrase acknowledges rhetoric's in-the-airness, the kind of currency Gronbeck intuited. When Everett Hunt, one of communication-rhetoric's early giants, attempted to document how World War II was changing the way people communicate, he located the biggest change in what he called "rhetorical mood." Rhetorical mood is the phrase Hunt uses to describe something detectable by what he calls an "instinctive sense," and something that can be intensified or changed by "the successful rhetorician."[40] Toward the end of his short reflection, Hunt uses the analogy of lighting to characterize the shift in mood: "the light seems to have changed now," he wrote, making postwar Americans less susceptible to glittery prose.[41]

In a 2008 tribute to Thomas Farrell's work, Maurice Charland, the rhetorical theorist who was among the first to elaborate a theory of constitutive rhetoric, culls from Farrell an emphasis on mood. Charland specifies that Farrell's conception of mood is more colloquial than grammatical, "as he [Farrell] writes of dispositions or attitudes toward aspects of common life."[42] Rhetorical moods bear a striking resemblance to what Lisa Corrigan calls "political moods," "unique emotional (dis)positions" that carry both "a chronology and a temporality."[43] Such moods constitute moments as much as they help form movements.

Felt time, time suffered, temporal *pathos*, matters for rhetoric, and for this tantalizing idea of rhetorical mood, because as Hunt notes, and as even Aristotle observes in the *Rhetoric*, feelings, *pathē*, are often the route to changing people's dispositions toward an issue or case. "The *pathē*," wrote Aristotle, "are those things through which by undergoing change, people come to differ in their judgments and which are accompanied by pain and pleasure, for example, anger, pity, fear and other such things and their opposites."[44] He goes on to divide his discussion (somewhat predictably) into three headings: the state of mind of someone with this feeling; the person, thing, or circumstance toward which they feel the feeling; and why they feel it: "for if we understood one or two of these but not all, it would be impossible to create" such feelings.[45] Two aspects of Aristotle's conception of feeling are worth noting. First, the *pathē* that matter most, as Aristotle presents them, are the dispositions (*hexeis*) of those who are essentially in power. And second, his commentary on the *pathē*, far from being the strictly psychological account of Citizens of Athens, as is the accepted reading, is profoundly rhetorical. In other words, the *pathē* can be both discerned and transmitted, even made. As Aristotle points out toward the end of his introductory chapter, a thorough understanding of, say, anger is necessary if one is to "create" it. The verb here, *empoiein*, installs a poetic making of feeling at the very core of rhetoric, or at least of the *Rhetoric*.

Carey and Corrigan both provide an important adjustment to Aristotle's conception of *pathē* when they locate feeling in time itself. Time, in effect, can carry and call forth, can create moods. Enabling that adjustment are perspectives—of Black women in Carey's case, and of Black civil rights activists in Corrigan's—that have for too long been marginalized. The testimonies of youth climate rhetors, too, conjure a mood of urgency composed of impatience, simmering anger, fear, or crushing anxiety.

As witnesses, these youth rhetors embody past, present, and future uniquely and simultaneously. Time and again, youth rhetors in these hearings make direct arguments about time and timing, and as they present those arguments, they also bring before their audience the feeling of having their futures taken from them, presenting their own versions of felt time and their own rhetorical mood. In doing so, youth rhetors show how rhetoric's time (*kairos*) and the clock, calendar time of *chronos* intensify each other. Their rhetoric of felt time intensifies the rhetorical mood of the two sympathetic hearing chairs. The hearing chairs, Congresswoman Kathy Castor and Congressman Bill Keating, repeatedly ask their audience to turn toward a future that they, the youth rhetors, face themselves. But the youth activists do so without losing sight of past injustices, especially those litigated in the civil rights era, like school segregation, which serve informally and formally as their precedent. Here is how the climate youth speakers are changing rhetoric: their wide view of time slides back and forth, their acute awareness of a future that hangs in the balance on display at every turn in their argument. Their rhetorical mood works in tandem with Thunberg's information-only approach, intensifying Thunberg's IPCC-report testimony by drawing out its implications for their own lives—by, that is, bearing witness to their futures.

Generational Listening

For the older, whiter members of Congress, the witnesses at these two hearings embody a future that can only be seen through the eyes and words of the young. The horizon for youth lies further out than that of the lawmakers. (The average age of representatives in the 116th [2019–2020] Congress, for example, was 57.6 years.[46]) In the April 2019 hearing, Select Committee Chair Kathy Castor weaves time and feeling into her opening statement. The different layers of time in that statement are worth dwelling on because they show how the climate crisis demands the widest possible temporal view, and they frame the witnesses in a way that sets the mood for the hearing, for a cross-generational listening. Having read the witnesses' written statements in preparation for the hearing, Congresswoman Castor presents the witnesses to her colleagues as, above all, generationally distinct from them.

First, then, there is generational time, the accumulation and division of years that separate age groups into generations. Congresswoman Castor presents the witnesses, in keeping with the hearing's title, as "generation climate," as a generation living without precedent in relation to climate. She does so by tracking lives of the younger generation with global temperature, a standard measure of the changing climate: "The last time global monthly temperatures were below average was in February of 1985. That means all of our witnesses and everyone who is 34 years or younger have grown up in a world that has been forever altered by climate change. In fact, this is the first Congress with Members who have grown up in the climate crisis. Six of our colleagues were born after that last below-average month."[47] Congresswoman Castor's generational frame allows her to present a division: those who, like their witnesses and their newly elected colleagues, have been born into crisis, a crisis not of their making, but one of their living. It's almost as if global temperatures themselves have brought forth a generation and a generational rhetorical mood.

This clutch of sentences also allows Congresswoman Castor to note that Congress is beginning to turn over, that members of the next generation are now their junior colleagues. Castor continues the generational argument in a way that explicitly presents her own generation as comparatively unaware of the crisis, using the colloquial phrase "back in the day" to describe the science curriculum followed in her school: "When I was in science class, back in the day, I didn't learn about how burning fossil fuels could change the climate, but students learn about that now."[48] She follows that observation by invoking the popular names for recent generations: "I am Gen X, but Millennials and Generation Z have grown up knowing we are in a climate crisis, and they are demanding that we address it."[49]

Congresswoman Castor's opening statement builds from a generational frame of innocence and awareness about the climate crisis to a frame of action—"urgent action, ambitious action" are her words.[50] Her oratorical crescendo includes a string of sentences that rely on repetition—in rhetoric, the figure is known as anaphora ("carrying back")—to establish the dominant time frame of the hearing, the urgent time for action, the now: "The time for rejecting climate science is over. The time for frustration and despair in the face of the climate crisis must end. This is a time for hope. This is a time for solutions. This is a time for all of us to come together—all generations, all political persuasions—for action. You all are rising to the occasion. We must rise with you."[51] Here the timeframe of her opening statement abruptly narrows to the now, the moment of decision and action.[52] But Congresswoman Castor does not pivot directly from the innocent, transformative, generational timeframe to the urgent, immediate one. Crucially, indeed, between the two timeframes, she pauses to attend to generational feeling:

> Seventy percent of young people in America say they worry about climate change. And based on the latest science from the administration's own National Climate Assessment and the Intergovernmental Panel on Climate Change, they have reason to worry. Seas are rising, snowpack is melting away, and in many parts of the country, droughts are getting worse. Hot, humid heat waves are becoming more intense. We are faced with more days where people cannot safely work or play outside. And higher temperatures mean that other pollutants, like ground-level ozone from car exhaust, will become even more damaging to our health.[53]

This paragraph weaves together facts and feeling; the prospects that inaction holds, according to the best science; and the feelings those prospects conjure, in this case worry. The paragraph sets the stage for the crescendo's temporal focus, where Castor speaks her wish to replace frustration and despair with hope. That the fact-feeling dyad follows so closely on the generational meditation reveals the committee chair's own hope for the hearing that day, for what the young witnesses would bring to her committee's deliberations.

Whereas Congresswoman Castor offered a statistical account of generational feeling in the spring hearing, in the fall hearing Congressman Keating asked outright:

> In all of your testimony, not just urgency came through when I was listening to what you had to say, but as someone from another generation listening, the last thing we would want for the generations to follow, for our children, grandchildren, and other people's children and grandchildren, is to hear in some of your remarks actual fear and anxiety being expressed.
>
> Could you, each of you, you can jump right in as you see fit, comment on what that is like? I think that that message should be heard by all of us, not just urgency, but what are we doing to the next generation? How are their lives being impacted by what we are not doing and what we are doing?[54]

This question repeats the word "urgency," the word Congressman Keating used in his initial transition to questions: "I thank all the witnesses for their testimony, and their urgency came through in all of your testimony."[55] His question loops back to the urgency point, which carries a kind of time, but he uses a contrastive construction, a "not just . . . but," when pivoting to feeling. He makes that pivot by invoking generations, specifically by the act of listening "as someone from another generation," listening across generations, listening for the future as manifest in the fear and anxiety young people are experiencing now. In this deliberative context, the generational timeframe intensifies feeling, to signal an urgency that often eludes the creeping pace of the climate crisis. And Congressman Keating's use of that timeframe to pose his question makes clear at least part of the reason these witnesses have been asked to testify in the first place: to help him help his colleagues see what their inaction is doing, to infuse the committee's work with the witnesses' fear and anxiety. But the youths' feelings, far from being separate from the urgency as Congressman Keating implies, actually carry the urgency, thereby constituting the urgent rhetorical mood.

Just as Congressman Keating wished in his first question for Greta Thunberg to expand on the importance of science, he wished for the other witnesses to expand on their feelings. Unlike Thunberg, who responded with five brief sentences, beginning with "Well, I do not see a reason to not listen to the science," and ending with "So, yes," Jamie Margolin took Keating's feeling question and answered for more than two minutes.[56] But before I turn to how Margolin and the other young witnesses respond to Keating's question, I will first spend time with the testimonies offered by them and by their peers at the spring hearing, the testimonies that effectively prompted Keating's question. Because as both Congresswoman Castor and Congressman Keating made clear in their opening statements, the young witnesses were there to galvanize their colleagues, to (as Castor put it during the question-and-answer period of the spring hearing) amplify "the fact that climate change is one of the most urgent, complex challenges humanity has ever faced."[57] And it is a crisis into which the witnesses were born.

Born into Crisis

One of the most compelling manifestations of felt time in both hearings is when witnesses speak of being born into this crisis. Two witnesses, Chris Suggs and Aji Piper, young Black men raised on opposite coasts, share what that looks and feels like to them: traumatic devastation and abuse, respectively. They bear witness to how long-term discrimination and abandonment feel, how they have come to know and recognize the effects of racist governmental policies in their (so far) short lives.

Kinston, North Carolina, where Suggs was born, is a town hollowed out by hurricanes. The poverty rate there ranges between 25 and 30 percent.[58] Here is what that looks like, according to Suggs: "Once a bustling community with a strong economy based on textiles and tobacco, today Kinston faces a lot of economic challenges. Before I was born, lots of industries left the area, and Hurricanes Fran and Floyd in the 1990s wiped out lots of business and damaged many of our neighborhoods."[59] He moves quickly to the effects of these storms: "This led to disinvestment, lack of community morale, and a significant population loss. Between 1990 and 2010, Kinston lost more than 16 percent of its population. East Kinston was particularly hit hard. A neighborhood there known as Lincoln City was wiped out by Hurricane Floyd in 1999, the year before I was born."[60] Suggs goes on to characterize the effects of Hurricane Floyd as both lasting and devastating—he uses the phrase "beyond devastating" twice: "To this day, there continues to be concentrated poverty and crime in my neighborhood, in part due to lots of abandoned structures, outdated and dense public housing, and lack of economic development."[61] Like the boarded-up, emptied-out structures Suggs sees daily, Kinston had been abandoned. Suggs was born into a community where each large storm exposes and intensifies its precarity. As he puts it, "poverty and hurricanes are deeply intertwined for us in eastern North Carolina."[62]

It gets worse. Suggs describes for the committee the devastating effects of Hurricane Matthew in 2016, and how, before the community could recover from Matthew, Hurricane Florence tore through the area in September 2018. Here is how Suggs recounts the whole of it: "In just my 18½ years on this earth, my community has experienced two 500-year floods on top of the floods after Hurricane Floyd in 1999. For these catastrophic events to happen at such a fast rate, a rate that my community can't recover from, is deeply alarming."[63] Suggs was born into the climate crisis, and he recounts how that looks and feels. He does so by offering a grave timeline of setbacks that outstrip recovery efforts, a chronicle of unrelenting devastation. When Suggs talks about the fast rate of floods, about two "500-year" floods in such a short timespan (three in twenty years, if you count Hurricane Floyd), he articulates his community's temporal precarity, a precarity that is, in his words, "deeply alarming." And it ought to be to members of the Select Committee, to members of Congress, to all of us. That, at base, is Suggs's powerful, if tacit, message. By framing his life with the rate of hurricanes, Suggs submerges his and his community's lives beneath the rising floodwaters, mixes them with the sediment of trauma and devastation the storms and floods leave behind. In his written statement, Suggs turns to the near future, to the coming year: "My community is still rebuilding from Florence, and from Matthew on top of that. And there's hesitation and fear in Kinston around what might happen this year."[64] Fear is, of course, the future-oriented feeling par excellence.[65] But by pairing fear with hesitation, Suggs adds a temporally infused action. Hesitation names an informed pause or delay; to hesitate is to enact uncertainty and doubt.[66]

Suggs then goes on to expand his witness, to lay before his audience the compounding effect of climate crisis, how it worsens the already harsh conditions experienced in places like Kinston:

> For me, the saddest thing about these recurring natural disasters that are exacerbated by climate change, is that the communities that are the most affected—like mine—are often the communities that have ALREADY been hit the hardest by all of society's other problems. You have poor, rural communities that are completely underwater or get cut off from their access to food, hospitals, and medical supplies. You have communities that rely heavily on the farming industry just devastated by these storms, causing farmers, migrant workers and their family to lose income while the farms are underwater. And you have predominantly poor communities, black communities and housing projects that were built in the flood plains—because those were the only places they were allowed—that become completely submerged. That's the story of Kinston, and much of eastern North Carolina.[67]

Notably, this passage is one of two from this hearing to be directly quoted in the Select Committee's June 2020 547-page report.[68] The felt time that Suggs presents here carries the feeling named at its outset, sadness—"the saddest thing"—while it lays out the conditions long endured by people in and around his community, conditions that were quite literally built into the community.

Suggs's written statement concludes by returning to the born-into-this frame: "I've never known a world that wasn't impacted by climate change, and it's time for that to change."[69] As the next chapter will discuss in the context of COVID-19 and temperature rise, climate change brings to the fore the idea of rate, which combines speed and time. It is crucial, therefore, to understand how rate works rhetorically: it conveys a mathematical magnitude that helps people glimpse, numerically, the severity of things. Here, Suggs uses the term of statistical probability to convey the effects of climate change: a 500-year storm is a storm with a severity that is likely to happen only once every 500 years. His discussion of the increased rate of storms in the context of Kinston and its hurricanes highlights the feeling of time, and the mood it creates over time, by showing the difficulty his community has recovering from storm after storm, flood after flood, in such a short timeframe. The result, in his account, is a communal disposition toward the future shaped by fear and hesitation. The members of Suggs's community, ever braced for the next disaster, live in a perpetual state of setback, each setback compounding the setbacks that preceded it. For them, the climate crisis is not a crisis of the future, but a crisis right now, one that turns each crisis into a preexisting condition (to borrow a term from medicine), and those preexisting conditions compound and magnify the crises to come. Suggs brought to the hearing that rhetorical mood—the devastation, fear, and hesitation—in all its poignant urgency. He stands as a witness to the past and present failure of all systems to protect Kinston, and the fear and hesitation he ascribes to the members of his community as they anticipate similar future failure. Suggs shows how felt time moves in an insidious feedback loop, creating the conditions for a mood to set in, a constant state of recoil. His testimony seeks to break the structured pattern of precarity by identifying it, calling it out.

If for Suggs and his community the climate crisis feels like constant fear and hesitation, like a long, relentless emergency, for Aji Piper, the third witness at the Select Committee hearing, the climate crisis feels like abuse. Piper's written statement compares the trauma of discriminatory policies to domestic abuse he suffered at the hands of his father: "As a young black man, I have grown up with the long-lasting consequences of unconstitutional discrimination from government-sanctioned and -engineered segregation. My childhood was shadowed by trauma from an abusive father. The trajectory of his life was formed in part by generational trauma of unlawful discrimination. Generations of black families have lived with the lasting legacy of government-sponsored racial discrimination. . . . Unconstitutional systemic government actions have long-lasting social consequences. Innocent children inherit those legacies."[70]

Like Suggs, Piper describes the feeling of being born into crisis. And Piper, in his written statement, names the upshot of that temporal feeling: harm of blameless children, of whom he is one. Whereas Suggs focuses on how destructive hurricanes exposed and deepened the impacts of discriminatory policies and their structural inequities, Piper amplifies what he calls "generational trauma" and the feeling of injustice. Indeed, the conclusion to his written statement lays bare those feelings, and its first two paragraphs merit repeating here as witness to the crisis and its future:

> Growing up with the looming threat of climate change has had last-ing impacts on my mental health. Thinking about the future has been a constant source of anxiety and depression for me. I have felt as if there is a pressure cooker boiling over inside of me. I can hardly fo-cus at times because I am overwhelmed with existential horror about the fate of the planet.
>
> I am a child of abuse. I know the feel of it on my skin and deep in my psyche. And what my government is doing to perpetuate indefi-nitely fossil fuel energy, and not take urgent comprehensive action to try to stop climate change, is a form of abuse on young people, who don't have the votes or the lobbying money to stop it.[71]

In the conclusion of his oral statement, Piper turns from chronicling deliberate discriminatory policies of the past to now-time: "The mo-ment is now. The moment is not for fear or incrementalism. This is the moment for heroism, for humanity, for standing with children around our country, standing with me."[72] His written statement issues impera-tives to the lawmakers: "Forget about being on the right side of history. If there even are history books, it will be because of the efforts that we are taking today. Be on the side of young people right now. Act as if our fundamental rights to life, liberty, property, and equal protection un-der the law are as important as yours, those who came before us, and those who will come after us."[73] The two ifs in this passage—"if there even are history books" and "act as if our . . . rights are as important as yours"—crack the crescendo. The first, a conditional, questions the very existence of any kind of future or potential legacy for the lawmak-ers, because their actions may well be too late, too little, or both. The second if, the one where Piper encourages committee members to "act as if" his generation's rights matter as much as theirs, calls for a change in behavior, albeit a fictional because perhaps idealistic change. He asks for equal treatment across generations, including those not yet born. Piper wants the legislators, that is, to enact the ideals they are bound to uphold.[74]

In the second hearing, Vic Barrett introduces himself as "a first-generation Garifuna American" whose "people are an afro-indigenous community originally from the island of St. Vincent in the Caribbean." Barrett begins his testimony by narrating how his ancestors in Hon-duras and Belize "organized our community and emancipated ourselves to protect our future as a people."⁷⁵ Noteworthy here is Barrett's use of the first-person plural, which complements his argument about the continuous cross-generational struggle that he and his people have endured. He expands his scope early on:

> It is not just me and my people in Honduras being harmed by climate change. Frontline communities around the country and around the world are already feeling the effects of the climate crisis from the dispossession of land to the grave public health threats that are disproportionately affecting myself and other young people.⁷⁶
>
> I, myself, have felt the consequences of climate change directly. Growing up in New York, I was impacted by the climate change-fueled Hurricane Sandy which left my family and school without power for days. I still experience grave anxiety about experiencing another climate-driven disaster like Superstorm Sandy and the harm that these storms will have on myself and my family.⁷⁷

Barrett's testimony offers yet another account of felt time, one that collapses his ancestors' past with the recent storm. That collapsing is necessary to account for his own anxious disposition and also to provide the widest timeframe possible. Like Piper, Barrett weaves the climate crisis into the long suffering of generations at the hands of governments (in this case "the British colonial power"), and like Suggs, he has experienced firsthand, and therefore fears, the devastation of a climate-intensified hurricane. And if the long timeframe of generational suffering compounded by the lingering effects of the specific storm disaster weren't enough, Barrett makes plain that the temporal feeling seeps into his daily routine, creating a dense mood of apprehension: "As someone who already struggles with anxiety and struggles with depression from my understanding of climate change and what I experience, watching our government knowingly perpetuate the climate crisis is extremely overwhelming. I wrestle with this anxiety every day from the moment that I wake up in the morning to the moment I fall asleep at night." In the next paragraph Barrett describes the future he fears the most, collapsing all the timeframes he has shared so far—the generational suffering, the single devasting storm, the government's ongoing inaction, the daily grappling—with their accumulating suffering into one horrifying picture: "If we keep going on with business as usual, both Honduras and New York, the places where my family and I are from, will forever be lost to the sea. That is one of my greatest fears, that climate change is going to take these places away from us."⁷⁸

Barrett ends his oral testimony with this crescendo, himself using anaphora: "I was born into a world in which my future and my past are uncertain, born into a world where my culture and inheritance are literally slipping into the sea, born into a world where my people are going extinct." The repetition of the words "born into" invokes Barrett's innocence, and the innocence of his people, leading into his final plea, cast in the frame of now: "Show children everywhere that you care about our future and the future of all generations to come. Now is your time to stand in solidarity with me and my co-plaintiffs, America's youth, and communities around the world to fight for a just future, free from catastrophic climate change. Thank you."[79]

The Feeling of "If"

If Suggs, Piper, and Barrett create a rhetorical mood by building to a description of how time in the climate crisis feels to them, Jamie Margolin leads with it. Here are the opening sentences of her statement: "My name is Jamie Margolin, and I am a 17-year-old climate justice activist from Seattle, Washington. I am missing a lot of school to be here. It is my senior year of high school. I have college applications deadlines looming over me, and to be honest, I have barely even started because I am too busy fighting to make sure that I am actually going to have the future I am applying to study for."[80] Margolin's opening sentences are filled with what she ought to be doing with her time—attending school, working on her college applications, worrying about deadlines for those applications and her school assignments. Margolin then considers how the time they are spending together is also something of a departure for the legislators: "You are here spending a few moments with me. But that is nothing compared to the hours that Members of Congress have spent with lobbyists from corporations that make billions of dollars off of the destruction of my generation's future."[81]

That disparity of time shared with fossil fuel industry lobbyists versus the future of younger generations feeds the demand Margolin makes next: "I want the entirety of Congress, in fact, the whole US Government, to remember the fear and despair that my generation lives with every day, and I want you to hold on to it."[82] In a few short moves, Margolin has laid before the lawmakers how it feels to her to have her time, her life, her future, so mishandled. She leads with the constant fear and despair, with deadlines that "loom" but are themselves overshadowed by the larger looming threat of the climate crisis. Margolin will later speak of exhaustion that she and her fellow youth climate activists feel: "We are exhausted because we have tried everything."[83]

But before she mentions that distinctly temporal feeling of exhaustion,

Margolin has words about her audience's own disposition toward the witnesses. Here, Margolin is responding to the common "patronizing" attitude she and her fellow activists have observed on the part of politicians. She writes about this attitude in her book, *Youth to Power: Your Voice and How to Use It*, a guide for aspiring youth activists: "Politicians always tell me, 'Keep up the good work; your generation is going to save the world!' They are both so right and so unbelievably wrong at the same time."[84] In the book, Margolin goes on to elaborate what power the youth voice can have, but she is careful to note what, exactly is so "unbelievably wrong" about this patronizing attitude.[85] What Margolin is getting at here is a rigid spectrum of attitudes that adults—especially upper- and middle-class white adults in the US—have toward children, and one of those is pride, an attitude that now leaves Margolin jaded. Here is what she says in her statement: "The fact that you are staring at a panel of young people testifying before you today pleading for a livable Earth should not fill you with pride. It should fill you with shame. Youth climate activism should not have to exist."[86] For Margolin, youth climate activists like her only do this work, spend this time, because the people before her (both temporally and spatially, "on the dais") did not do their jobs. In *Youth to Power*, she calls it "unacceptable for leaders to put the burden of the problems they created and have the power to fix on our shoulders."[87] When Margolin elaborates this point, she encapsulates perfectly the temporal bind that youth climate rhetors find themselves in: "If we wait until Generation Z is old enough to be in power, it will be far too late to address the crisis. In fact, we are already deadly late on acting on the climate crisis."[88] To the lawmakers at the hearing, she applies the direct heat of accusation: "You are promising me lies. Everyone who will walk up to me after this testimony saying that I have such a bright future ahead of me will be lying to my face."[89] Margolin has come to meet this patronizing disposition with a disposition of refusal. In *Youth to Power*, Margolin advises her readers to defy the expectations that adults may have of them and to "Be the most patient, mature, and polite person in the room."[90] At the same time, she writes: "There are some situations where you just gotta say, 'Screw it,' and do you. Freak out, get mad, yell when it's called for and appropriate and the issue is super serious. Use your best judgment. Being exactly what the world thinks you are—angry and loud—is just fine when you're calling out a politician in public for doing something horrific."[91] The early part of her statement at the crowded hearing reads like just one of those times for Margolin. But her measured delivery does little to indicate that Margolin is freaking out.

To return to Margolin's timeframe, whereas every sentence in her opener refers to time in some way, the "looming" application deadlines and the anxiety they bring cast a mood over her statement. The deadlines that loom, like clouds in the Pacific Northwest, heavy and dark, are perhaps less consequential than the deadline the IPCC has set for action. Margolin, whose movement is called Zero Hour, is well aware that time to act is dwindling. As her statement moves to discussing the larger deadline, she shifts into the future perfect tense, much like the plaque now installed at the site of the former Icelandic glacier discussed in chapter 2; only in this case, the futures faced by the audience members and the witnesses are already unequal. Here is Margolin:

> By 2030, we will know if we have created the political climate that will have allowed us to salvage life on Earth or if we acted too late. By then, we must be well on the path to climate recovery.
>
> But this must start today. In fact, it should have started yesterday.
>
> By 2030, I will be old enough to run for Congress and be seated right where you guys are sitting right now. By then, we need to have already achieved net zero greenhouse gas emissions and be rapidly on the path to climate recovery.
>
> I cannot wait until I am sitting in your seats to change the climate crisis. You have to use the seats that you have now because by the time I get there, it is going to be way too late.[92]

Her "I cannot wait" is not the same as the phrase young people use to describe an excited, eager mood, the one of anticipating an event—a vacation or a concert, say. Instead, it names a clear limit and performs a version of what Carey calls rhetorical impatience: waiting is out of the question. Such is the nature of time and the climate crisis. Referring to Thunberg, who testified right before her, Margolin returns to the deadline language: "As Greta mentioned, the Intergovernmental Panel on Climate Change report that we only have a few months left in order to create the massive political shift needed to transition our world to an entirely renewable energy economy. This needs to happen within the next 10 years, which is our deadline to save life as we know it."[93] The 2030 deadline and the question of action together guide the conditional and future tenses Margolin uses—a mix of simple future tense ("we will know") and future perfect tense ("if we have created the political climate that *will have allowed us to salvage*"; "we *need to have already achieved* net zero greenhouse gas emissions"). These statements, that is, follow a "by that point" logic, and Margolin delivers them without reference to feeling—the deadline itself carries the pressure, the urgency, of which she speaks.

Margolin makes time feel urgent by invoking the 2030 deadline, casting the legislative body into the future.[94] With the anaphoric repetition of the short word "by"—"by 2030" repeats twice, both times carried forward in her remarks with the phrase "by then"—Margolin performs the future memory work I discussed in chapter 2. Recall here the arresting lines from the Okjökull memorial plaque, which, like this section of Margolin's statement, focus on curtailing carbon emissions: "This monument is to acknowledge that we know what is happening and what needs to be done. Only you know if we did it." As I noted, the monument conjures future witnesses by addressing them directly, granting them the knowledge of past action or inaction, carried by the conditional, "if." Margolin's testimony, though, calls forth her future self as one of many judges. She and her panelists are the "we" who in 2030 "will know" if the "political climate" has changed. The present audience effectively recedes into the uncertain future they helped create. In this way, the youth climate activists, as Congressman Keating indicated in his opening statement, bear witness to both the present and the future.[95] The climate activists, members of their generation, and generations to follow will know if current leaders did what needs to be done. Indeed, they already have an inkling.

I want to note here the choice of the phrase "political climate" to refer to the state of US politics in 2019. With this phrase, Margolin names the most daunting obstacle to addressing the climate crisis. She elaborates:

> The good news is that experts agree that there are multiple pathways to decarbonize the United States energy system and that doing so is both technologically and economically viable and beneficial.
>
> The most frustrating thing is that the U.S. Government cannot even begin to imagine the massive political shift that has to happen in order for us to solve this issue. The politics just has not been invented yet.
>
> Solving the climate crisis goes against everything that our Country was, unfortunately, built on, colonialism, slavery, and natural resource extraction. This is why the youth are calling for a new era altogether.[96]

At this point in her statement, Margolin's characterization of the problem has been delivered in even tones. But here, voice rising, Margolin names frustration in the transition from the "good news"—the fact that solutions or pathways to zero emissions exist—to, effectively, the bad news: those pathways are being blocked by many of the people in front of her, and there is no longer time to wait for those people to leave office.

Margolin has recast the climate crisis as largely a rhetorical, temporal problem. That is not to say that science is not real—far from it; rather, she means that the lawmakers, as a collective, remain unmoved by the science. This is where youth rhetors come in, and it is why Thunberg's stark testimony ("I don't want you to listen to me, I want you to listen to science") needs the other witnesses to provide the mood, to elaborate the felt time on which her testimony rests. Put simply, change this radical requires a jolt that information has trouble providing. This is the task of feeling: people need to be moved to act. Even Aristotle knew that. Congressman Keating seems to know it as well: he and Congresswoman Castor need the youth rhetors to help them and their colleagues see the bleak, uncertain futures the youths see, to know the injustices they know, to feel the urgency they feel. Indeed, when Congressman Keating explicitly tries to further draw out the witnesses' feelings, Margolin is the first to respond:

> For me, it has really been affecting because, similar to Vic, I already have underlying issues of, anxiety, and it is just really hard to grow up in a world full of "ifs."
>
> You know, I do not think a lot of people in Congress understand the conversations that are happening in everyday American high schools, but we are constantly asked: Prepare for your future, study for your future, do this for your future. But our world is full of "ifs."
>
> I will be talking to my best friend, and she will say, "Yes, you know, I really want to see this natural place sometime *if* it is going to still be around. I really want to study to be this *if* that is still going to be a possibility."
>
> And it is just, this constant looming uncertainty. And it is a weird form of nihilism and weird just fear that has been existing in my generation where kids are joking, what is even, is the point? The world is ending. What are we studying for? What are we doing?
>
> And it is this kind of depression, it is this fear that is not just among me or my panelists here but everyone, and that anxiety is something that no child should ever have to fear.[97]

Here, Margolin's depiction of her everyday life—her "world full of 'ifs,'" like Ari Piper's "if there even are history books"—conveys the pervasiveness, the all-consuming nature of the feeling of a threatened future. Margolin's description may be most productively read as that of a rhetorical mood, which, as Corrigan and Charland observe, constitutes political interventions, gives these interventions a *charge*. In describing the mood and how it inflects conversations, punctuating wishes and dreams with question marks and brackets, Margolin offers more than an airing of grievances. She offers relentless critique, witness to inter-generational injustice. Her answer to Keating's question speeds up as her pitch rises:

> Because if you think about it, if you go back to what is the purpose of a parent down to just the biological purpose, it is to give their child the best future and the best life that they can possibly have and the supposed American Dream is to make sure that children have a better future than the adults.
>
> But right now, it is, some members of government and some cor-porations are actively pointing a gun to children's futures and actively making it worse, actively going out of their way to support corpora-tions and poison us and destroy our future.
>
> And that is horrifying, and it feels like a betrayal, it is like a knife to the heart, to know that people who have kids, they will go around in these campaign ads and they will be, holding these babies, "Oh, you should vote for me, look at me interacting with a small child," while they actively poison and choose their wallets over their children.
>
> So it is very devastating and scary, but also it feels like we have been betrayed.[98]

If Keating wanted feeling, Margolin brought it. She did so by provid-ing a catalogue of interrelated and escalating feelings: anxiety, depres-sion, fear, the pain of betrayal, devastation. These feelings, as Margo-lin presents them, are interwoven with facts, what she knows and sees, and wound with time, what Martin Luther King Jr. famously called "the fierce urgency of now."[99] The feelings Margolin describes for the lawmakers, that is, are intensified by the now fast-approaching dead-line—by time's passing. Margolin uses the language of deadly weap-ons—a gun, "a knife to the heart"—and Aji Piper in the spring hearing used the pressure cooker image.[100] Both analogies carry the threat, and therefore convey the fear, of a dramatic, even violent end.

In the question-and-answer session, a telling exchange happens with Congressman Garret Graves, the Select Committee's ranking member from Louisiana. Mr. Graves poses the following question to Greta Thunberg, again cast in terms feelings: "If you are sailing across the ocean and you are picking up trash along the way, and every one piece of trash that you pick up there is a boat right next to you dump-ing out five pieces, how would that make you feel?"[101] Congressman Graves here offers a thinly disguised "metaphor" (his word) for the US and China. After back and forth, Graves yields back to Congresswoman

Castor, who sums up the position held by Congressman Graves and others (including then-President Donald Trump) in this way: "Some people say that the United States should not dramatically reduce our emissions because China and other countries are not doing enough. I would like to have your view on that and have each of the witnesses comment on that briefly."[102] Benji Backer, the sole conservative activist and only college student on the panel, takes the question first, and essentially rejects the premise, adding "I do not think that that is a reasonable excuse."[103] This is the point in the hearing when Jamie Margolin, to use the language from *Youth to Power*, freaks out. When it is her turn to speak, she reverses hearing protocol to ask members of the committee a question. Her fierce intervention is worth quoting at length, because it serves as something of a climax to the whole hearing and stands in marked contrast to Benji Backer's debate-like approach and Greta Thunberg's act-on-the-science approach. It carries forward both the mood she sketched at Congressman Keating's request and the shaming of the lawmakers in her opening statement:

> I have a question. When your children ask you, "Did you do absolutely everything in your power to stop the climate crisis, when the storms are getting worse and we are seeing all the effects of the climate crisis?" when they ask you, "Did you do everything?" can you really look them in the eye and say, "No, sorry, I could not do anything because that country over there did not do anything, so if they are not going to do it, then I am not"?
>
> That is shameful, and that is cowardly, and there is no excuse to not take action, to not improve as much as we can in the United States.
>
> And how can we call ourselves the city on a hill or be an example for the world if we are going to be coward and hide behind waiting for other people, saying that, "I am not going to do this because they did not"?
>
> I want you to think about this is all about being able to look your children in the eye and say, "I did absolutely everything I could for you. I know that we are up against a lot of pressure. I know that time is running out, but, honey," however you call your kids, "I did everything I could."
>
> And so I just do not understand as a parent how can you look your kid in the eye and say, "There is this impending crisis, everything is at stake, but I stood back and I did not really do anything. I did not take action. I did not act like it was an emergency because our neighbors over there weren't doing it, so I am just not going to." How can you tell your children that?[104]

Margolin's imagined family scenario performs the enchainment of Magnason (considered in chapter 2), but instead of a chain strengthened and formed by love, the chain she imagines is broken by betrayal. Like Aji Piper in the first hearing, Margolin is effectively asking the committee members to "act as if" they care about their children, the world, the future. Margolin's pitch, along with her accusation of cowardice, together perform the frustration she calmly spoke about in her opening statement. She asks them to imagine their future accountability, if not to the American people, then to their own grandchildren. In a way, Margolin's scenario is the mirror image of the loving generational enchainment Magnason encourages young people to imagine. Here, instead of familial affection, Margolin proposes betrayal and familial shame. Margolin effectively enacts Piper's pressure cooker analogy. Her rupture offers a glimpse of a new form of witnessing, one that performs the urgency that runs just beneath the surface of felt time. None of the enactments of impatience—dismissal, indignant agency, and repossession—that Carey attributes to Black women are available to BIPOC youth climate activists. Their agency is diminished and smoothed over with the patronizing narratives of pride that make Margolin seethe. Seeing no capacity for shame at present, Margolin decides to take her audience to the future to glimpse it there, and in doing so she flips the rhetorical table. Margolin's response digs into her role as witness for the future and finds in witnessing's traditions of morality and justice a shame-filled future for the lawmakers by conjuring for the congressional committees an intimate reckoning. Mood in this instance slips into temper.

Conclusion: Mood

The witnesses whose testimonies I have examined in these hearings brought a distant future closer, held it up for the lawmakers to see, gave it presence, cast it in a mood. Greta Thunberg gave presence to that future by passing along the science on the matter, presenting a report that explains what will happen if humans fail to keep temperature rise below 1.5°C. If, as the rhetoric of science scholar Lynda Walsh [Olman] argues, scientists have assumed the role of prophet, peering into the future and sharing what they see in times of crisis, then in the September 18 hearing, Thunberg's testimony made her into the prophets' prophet.[105] That Thunberg thought it was necessary for her to amplify climate scientists before the congressional representatives reflects her own conception of American political inaction as guided by her public interactions with then-President Donald Trump, who was overtly hostile to climate science, rolling back environmental regulations and with-drawing the United States from the Paris climate accord.[106]

The import of Thunberg's testimony lies in its work as a backdrop, as the basis for the felt time presented by her American peers. The contrast between Thunberg's testimony and that of the other youth witnesses exposes the limits of information sharing when it comes to the climate crisis, when it comes to American congressional politics, and in this case, when it comes to both. Thunberg's testimony, that is, gives the lie to the "information weighing" function of these hearings. Feelings, after all, can weigh too, at times so much that a crack forms.

The BIPOC and LGBTQ youth witnesses in these two congres-sional hearings bear witness to the future. They do so by drawing out time's quality through the suffering they and their communities have endured and the variegated feelings that suffering brings about. What these youth witnesses put forward is a version of rhetorical change that on one hand dates back to Aristotle's account of rhetoric as disposi-tions meeting dispositions. But the accumulation of injustices, some-thing that Aristotle could not from his privileged perspective fathom, creates dispositional friction, a negative charge that feeds the sense of urgency *as a sense*, a rhetorical mood. The youth rhetors in these hear-ings, that is, present the feeling of time itself, the accretion of injustices, the mounting setbacks. Their narration of felt time slides easily between past, present, and future, and by necessity collapses into one big tangle of feeling, a rhetorical mood of bare exhaustion and outrage, a view of the future teetering on the tiny word "if." Together, that is, the witnesses picture a bleakness that is difficult for many to fathom, but fathom it these witnesses do, because they feel it intensely. They live it.

Thanks to the efforts of these youth and others like them (Alexandria Villaseñor, Xiye Bastida, Xiuhtezcatl Martinez), "climate emergency" was named by Oxford Languages the word of the year for 2019.[107] That focus would, of course, all change early the following year with the ar-rival of a global pandemic, the subject of chapter 4.

CHAPTER 4

Learning Curves

COVID-19, CLIMATE CHANGE, AND
MATHEMATICAL MAGNITUDE

The virus was like a revolution in the brain, a brand-new argument.

ANDREW O'HAGAN, "Keepsakes"[1]

I was in the earliest stages of writing this book when word began to spread about a novel coronavirus emerging in Wuhan, China, a city of eleven million people. As we now know, the virus itself arrived in the US not long after word of the virus arrived—it's just that words were more discernible than the virus at that point. In January 2020, the virus was circulating in places far from Wuhan, places like the Lombardy region of Italy, where the first cases of COVID-19, the respiratory illness caused by the new virus, are now known to have appeared in Europe.[2] Snohomish County, Washington, saw the first known case in the US when an infected traveler arrived from Wuhan in mid-January, and a February 6 death in Santa Clara County, California was later determined by a coroner to have been COVID-19 related.[3]

The memory remains tender (and is likely to remain so): across the world, region after region shut down, schools closed, and economies flickered and dimmed, confining, almost overnight, education, office work, and broadcasting to basements, guest bedrooms, home offices, and bathrooms, even as essential workers continued to report for jobs that had become more demanding and dangerous. Other jobs—about 14 million in the US—evaporated. No life, that is, went unaltered by the pandemic. Phrases such as "unprecedented times" and "navigating the unknown" became commonplace.[4]

With astounding suddenness, COVID-19 stretched the global public's imaginative capacities, and it did so in ways that linked the pandemic to the global climate crisis. First, because of the drastic distancing measures needed to slow the spread of infection, many of the hardest-hit areas saw measurable drops in carbon emissions.[5] Second, the pandemic laid bare the serious consequences of excessive development and the habitat destruction and environmental degradation that results. These changes alter the planet's biodiversity and create what epidemiologists call vector ecologies, where diseases can jump from non-human to human animals.[6] The novel coronavirus, by most accounts, originated zoonotically.[7]

But perhaps the most vivid, complex, and nearly instantaneous stretching of the imagination, one that helped also to convey the magnitude of the climate crisis, was this: with the help of epidemiologists, public health advisers, and a few political leaders, the virus taught people how to see, think, and talk in curves. And much as the coronavirus appears to have leapt from species to species, the curve—specifically the curve used to forecast and understand epidemics known as the epi curve—leapt from the public health crisis to the climate crisis. That leap is the focus of this chapter.

The global campaign to "flatten the curve" of COVID-19 made visible and more accessible the complex mathematics of exponential growth; as such, the pandemic boosted public capacity to imagine that growth with its picture of steep and accelerating spread, unmanageable with no intervention. In effect, one unprecedented crisis (the global pandemic) provided a crucial means of intensification for another (the climate crisis) in the form of a visible curve, along with a specific, distinctive, sloganized mitigating action—to flatten that curve. A version of this curve is shown in figure 4.1.[8] That curve would soon be recognized by a global public.[9] Many would come to possess acute (which is to say felt) awareness of what the curve portends. In the months and years that followed, the curve would proliferate and, with time, begin to undulate, sometimes wildly, with infections rising and falling in waves.[10]

The curve graphic instantly became iconic, circulating around the globe in newspapers and on social media. That widespread circulation enabled the spread of what Jordynn Jack has previously called a "pedagogy of sight," allowing people to glimpse the profound magnitude of the pandemic.[11] The tool of visibility, in this case, is a generic form of a data visualization known as an epi curve, a graph of a disease outbreak that shows the number of infections over time. Generally, an epi curve can provide, at a glance, an overall sense of the pattern and spread of an outbreak, as well as what the CDC calls a "general sense of the outbreak's magnitude."[12] Elaborating the rhetorical contours of that "general sense" is part of the aim of this chapter. The graphic in question presents two epi curves. The first, "tall" curve shows infections increasing at a dangerous rate with no intervention; and the second curve, much

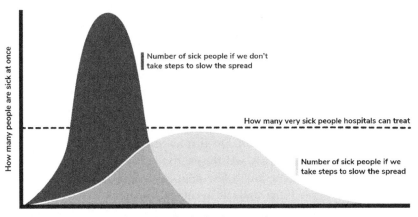

How many people are sick at once

Number of sick people if we don't take steps to slow the spread

How many very sick people hospitals can treat

Number of sick people if we take steps to slow the spread

How long has the virus been spreading

Adapted from the CDC   copyright University of Michigan

4.1 Epi curves. Courtesy of Dr. Howard Markel. © University of Michigan.

flatter and wider by comparison, shows case rates spread over time through nonpharmaceutical interventions such as physical distancing, school closures, and close-contact quarantines. These are, in the rhetorical logic of the diagram, two possible futures, meant to be compared.

As I discussed in the introduction, the unprecedented upends ideas of smooth, linear progress and the easy reliance on events from the past as guides for deliberating about futures. The epidemic curves in this graph visually depict two futures that require and shape imagination as the bridge from prediction to action.[13] But in an unprecedented situation, they do more than that. Specifically, as I will discuss in this chapter, these curves depict a mathematical magnitude that takes on and carries feeling as the numbers took on the magnitude of illness and death witnessed firsthand, by word of mouth, and by means of media reports. In this way, the pandemic's first curve, in the first part of 2020, carried particularly intense feeling—carried it right into the climate crisis. In short, the unprecedented global crisis of climate change got a rhetorical assist from an unprecedented global health crisis. This assist is evident in the temporary graphical merger of the epi curve with curves commonly circulated by climate scientists. In part, that assist was possible because unlike the climate crisis, the pandemic unfolded right before people's eyes, remaking their worlds.

The story of the data visualization's movement—the design changes it underwent to help people better see (and feel) the urgent need for action, and its quick, brief move into climate discourse—shows how a future-oriented visual object can gather intensities, stretch imaginations, and make accessible and transferable the relative mathematical magnitude of what people are facing. In short, the pandemic's curve provided a salient, visual means of intensification for the climate crisis. By detailing how that is so, this chapter contributes to scholarly conversations that focus on the relationship between rhetoric and mathematics, overlaying them with an area of study—data visualization—that has been worked over by scholars in a range of fields.[14] As most graphic designers, STS (science and technology studies) scholars, technical communication scholars, visual communication scholars, and economists already know, a good graphic can alter perspectives by showing information compellingly and accessibly.[15]

Indeed, STS scholars, technical communication scholars, and visual communication scholars have long studied how visuals argue.[16] Lynda Walsh, a rhetoric scholar working in STS and technical communication, helpfully organizes visual rhetoric approaches to climate change into three types: the classical, which foregrounds Aristotelian concepts like the enthymeme to think about how visual arguments allow shortcut associations; the semiotic, which focuses on verbal and symbolic sign use; and the critical, which draws from philosophical work by scholars like Foucault and Marx.[17] Walsh shows how these approaches might play out in an analysis of the now iconic hockey-stick curve that Michael Mann and his collaborators created to capture a predicted rise in global temperatures. It maps temperatures for the Northern Hemisphere using proxy data for the past 1,000 years. Mann's curve, as I will show later, gets incorporated with the epi curve graphic as it moves into climate discourse.

My approach in this chapter does not easily fit into Walsh's categories, in part because the version of rhetoric I am developing in this book—rhetoric as an art of intensification—does not align perfectly with argumentation (Walsh's category). Nor does my approach align with cognitive approaches to affective imagery. Intensification, that is, is a more oblique and not quite so straightforward framework for understanding rhetoric than is argumentation. Nor do I apply concepts and perspectives in a more or less linear way, as do the approaches Walsh describes.[18] If I were to do so, this chapter would probably find in the story of the movement of the epi curve from one crisis to another an open-and-shut case of *kairos*, that idea of opportune time discussed (and refined) in chapter 3. Such a story would go something like this: climate scientists, activists, and writers "seized the moment" of the pandemic. That story is accurate as far as it goes. But telling it in that way does not go far enough, for it affirms an old conceit about how exigencies rise and fall and how they must be created anew. As such, that account would reify the conceptual starting position, working in one direction: rhetoric helps show how technical graphics adapt to an occasion.

In contrast, a "following" approach remains alive to what the epi curve's different permutations teach us about rhetoric and precedent or lack thereof, introducing a new way to think about relations between rhetoric and mathematics. The story that emerges from the "leap" of the graphic from one crisis to another is a story of borrowed precedence achieved through circulation. It therefore aligns with the ecological, circulation approach taken by Laurie Gries, who tracks the rhetorical life of a particular image across time and space. In Gries's work, rhetoric becomes an "unfolding," its objects transformed as they move and circulate. She shows how "images become rhetorical as they produce and participate in events that are distributed, divergent, and unpredictable."[19] The story of the epi curve, a graphic used to teach mathematical principles of compound growth in a quickly unfolding pandemic, therefore becomes a story of a graphic gathering intensities along the way, pulling into it the rising feeling of what it is like to have a future pried away—what chapter 3 called "felt time." The brief cross-crisis life of the epi curve shows, that is, a graphic's "capacity to mobilize *magnitude*," a point that Finnegan makes with respect to photography.[20] The viral graphic, as I show, redesigned to picture magnitude at the outset, when circulated in a rapidly emerging crisis, layers mathematical magnitude with felt magnitude, and that augmented magnitude carries right into another crisis.[21] Following the graphic's move into the climate crisis brings forth three insights about rhetoric: 1) it showcases a dimension of magnitude that scholars have yet to consider: the eye-widening magnitude of exponential growth; 2) in doing so, it draws mathematics, the discipline ostensibly least concerned with feeling, to the center of a felt rhetoric; and 3) it presents a new way for STS scholars, and technical communicators in particular, to think about rhetorical effects by enlarging a point about artifactual appropriation made by the rhetoric scholars Robert Hariman and John Lucaites. Following the newly iconic epi curve and documenting the changes it underwent also, then, documents effects of rhetoric itself.[22]

To account for the multiple magnitudes pictured and accrued by the graphic, I want to first describe how the graphic in question was redesigned in ways that allowed it to take hold in the early days of the epidemic, in the days before and just after March 11, when the World Health Organization declared COVID-19 a global pandemic. Next, I follow the graphic's leap into the climate crisis. To that crisis the graphic carried with it neither true numbers nor perfect analogies. Instead, it carried the astonishing magnitude of exponential growth and its newly urgent lessons about the need to take action before the crisis is out of hand. The chapter then considers the change evident in climate discourse on the level of both image and word, including the replacement of the verb "bending" with the verb "flattening" when speaking of curves, and then the visual reframing of climate science's most iconic curves, the hockey-stick curve and the Keeling curve (the graph showing the steady rise of carbon dioxide in the atmosphere). All told, the early rhetorical life of the flatten-the-curve graphic and slogan provided a means of intensification, one that altered the rhetorical landscape for public health and also, briefly, for the climate crisis. Those early months of the COVID-19 pandemic also demonstrated, in a compressed timeframe, the premise of this book: that an unprecedented crisis can sweep rhetoric, too, into its large-scale change, altering rhetorical practice, calling for and constituting—sometimes on the fly—newly salient resources with which to intensify communication. The graphic and its slogan, in short, offer a window on how numbers can feel.[23]

Spreading Word and Image

On March 14, 2020, as COVID-19 was rapidly spreading around the globe, New Zealand's prime minister, Jacinda Ardern, held a press conference in Auckland to announce what she called "the widest-ranging and toughest border restrictions of any country in the world."[24] Her opening remarks carried a message of unity, unflinching resolve, and speed: "We must go hard, and we must go early. We must do everything we can to protect the health of New Zealanders. That is exactly why to tackle this global pandemic, Cabinet made far-reaching and unprecedented decisions today, because these are unprecedented circumstances."[25] In her roughly thirteen-minute address, Ardern laid out the new restrictions, implored the people of her country not to make a run on grocery stores, and advised them not to panic, but to be sure to wash their hands frequently. Ardern repeated the word "unprecedented" to frame the severity of the measures she was about to announce. That repetition served as a pretext to the deliberative (future-facing) portion of her speech. This is when she exhorted New Zealanders to join together to flatten the curve.

But before she made any of these points, Ardern concentrated on the matter of managing the healthcare system: "The key," she said, slowing her pace as she worked her way to the end of the sentence, "continues to be leaving our hospital system to those who need it most." After a pause, she went on: "All of this points to one strategy: spread the cases, and ultimately flatten the curve. That is how we ensure that health services are

4.2 Ardern explains how New Zealand will flatten the curve. By permission of *Stuff*. Image credit: Rick Wilson.

there for those who need them most."[26] At this point, less than two minutes into her remarks, she picked up a piece of paper and showed it to the people and cameras before her (see figure 4.2). On the paper were two comic-style graphic frames, one on top of the other (figure 4.3). These frames depicted the epidemic curve to which Ardern referred, showing a rise of infections over time. The first, in a darker shade, depicted the course of the disease's spread with no measures taken to slow it. Here the case rate rises sharply and overwhelms the hospital system (the capacity for which is represented by a straight dotted line). The second curve, in a lighter shade, rises more gradually over a longer period of time and remains below the dotted line.

Ardern had this to say about the graphics she held in her hand:

But the scale of how many cases we get and how fast we get them is something we should do as much as we can to slow. That is how we

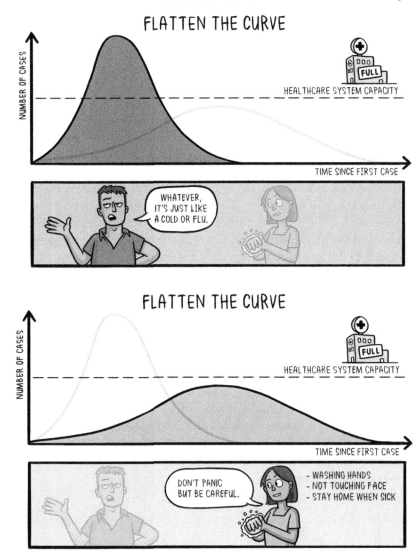

4.3 "Flatten the Curve." Credit: Siouxsie Wiles and Toby Morris.

ensure health services are there for those who need them most. And that is essentially what this graph by [microbiologist and science adviser] Siouxsie Wiles illustrates. In the case of a pandemic which hits a country hard and fast you see the curve of cases rises steeply. This of course puts pressure on health systems; it means it's more difficult to treat people in the place that they need it most. The goal is to ultimately flatten the curve. That doesn't mean you have no cases. It means the pace at which you receive them is at a rate such that you can make sure people are cared for in the places they need it most, be it moderate symptoms at home or be that in critical care if required. Ultimately, New Zealand wants to flatten the curve.[27]

As a visual anchor for her remarks, the illustration of the two epidemic curves allowed Ardern to help her audience see what she and cabinet members saw: one future where their country allows unmitigated spread, and another where it enacts serious but reasonable measures to slow that spread. The data visualizations helped Ardern to convey to New Zealanders the scale or magnitude of the situation, thereby helping her justify the unprecedented severity of the measures she was announcing. The images, in other words, supported, and indeed illustrated, Ardern's main message about the urgent need for mitigation. And they did so by presenting, one after the other, two future scenarios, charted in relative mathematical terms, one spelling havoc and the other promising calm, accessible treatment for those who need it.

Ardern's lesson on flattening the curve was one of the earliest and most direct accounts of why the pandemic required intervention, but hers was by no means the only one being delivered around the world. In fact, the graphic she held up—the two-curve infographic designed by Wiles, a microbiologist at the University of Auckland, in collaboration Toby Morris, an illustrator working in Auckland—was based on a more rudimentary version of the line graph shown in figure 4.4. That version had been circulating online since February 28, when Drew Harris, a population health analyst at Jefferson University in Philadelphia, created and posted it on Twitter to help convey the risk of overwhelming hospital systems.[28]

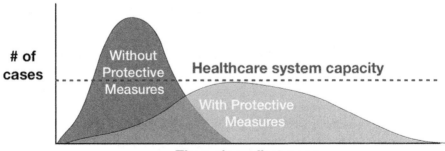

4.4 Drew Harris's version of the epidemic curves with the crucial, clarifying dotted line. Credit: Drew Harris.

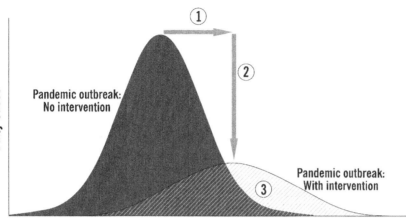

Goals of Community Mitigation

① Delay outbreak peak
② Decompress peak burden on hospitals / infrastructure
③ Diminish overall cases and health impacts

①

②

Pandemic outbreak:
No intervention

Pandemic outbreak:
With intervention

③

Daily Cases

Days Since First Case

4.5 Draft Zero: Goals of Community Mitigation. Credit: Centers for Disease Control and Prevention.

The original context for Harris's graphic was his work with hospital staff in a pandemic preparedness program.[29] He needed to help them understand the importance of managing a surging epidemic. Harris's is a modified version of a graphic included in a 2007 CDC pandemic preparedness manual (figure 4.5).[30] The CDC's manual, as its main title indicates, offers "Interim Pre-Pandemic Planning Guidance."[31] Its express purpose—"to provide interim planning guidance for state, territorial, tribal, and local communities"—dictates its focus on what is known in public health as "nonpharmaceutical interventions," because the scenarios therein are designed to buy time and reduce harm while pharmaceutical interventions (e.g., effective vaccines or treatment regimens) are identified or developed. Harris found that the people in his training program were not grasping the significance of the surge, so he added the dotted line to indicate the system's capacity to handle infections at such a scale.

The CDC image included a numbered key in a text box centered at the top to indicate how intervention can change the outcomes pictured by the two curves. Placed between two others, the second goal, to "decompress peak burden on hospital/infrastructure," loses emphasis simply through its location, and the key itself thus undermines the visual effectiveness of the graphic. Here is how the figure is referred to in the guide's text: "Delaying a rapid upswing of cases and lowering the epidemic peak to the extent possible would allow a better match between the number of ill persons requiring hospitalization and the nation's capacity to provide medical care for such people."[32] Little wonder the significance of the two curves did not quite sink in for Harris's trainees; their eyes were being asked by the graphic to move between the three numbered goals and the work of the curves, comparing case rates in an epidemic of unknown scale while also trying to imagine for themselves the impacts of each course of action. Importantly, this graph is conceptual, created to convey relative infection numbers of a future epidemic (or pandemic)—hence the absence of values on either axis.[33] The dotted line Harris added to his version, showing the case levels that hospitals can manage, cuts across both curves, both futures, with radiant clarity, revealing "what was at stake" (in Harris's words) in mitigating an epidemic surge.[34] The capacity line, that is, reduces the cognitive load of the visualization by omitting the key and instead placing labels directly on the graphic, calling attention to the importance of hospital capacity. The emphasis is provided by the line, or in Ardern's case, the slower pace she used to explain the importance of preserving hospital resources.[35] Harris's simple tweaks converted the already important CDC graphic into a visually potent pedagogical tool, one that places before the eyes the magnitude of the problem by showing the urgency of what his version of the graphic labels "protective measures." These changes made the graphic at once more powerful and more accessible, and that accessibility broadened the image's audience from local leaders on pandemic planning committees to hospital staff—those whose working conditions are depicted by the line—and, in late February 2020, as COVID-19's spread in the US became apparent, to the broad public, initially via Twitter. Harris's amended graphic is as strong an example as any of how appropriation and circulation can alter audience, and as such it provides artifactual evidence of rhetorical effects.

The Wiles and Morris version of the image shown by Ardern goes further, separating out the two line-graphed scenarios and including at the bottom of each scenario an illustration of a human figure practicing behaviors keyed to rising or falling case rates, respectively. This graphic also took the form of a single .gif, a one-second animation in which one image fades into the other: the curve flattens before the eyes, and as the curve flattens, the human figure changes from a figure presenting as a man voicing skepticism ("Whatever, it's just like a cold or flu") to one presenting as a woman, hands in a lather, whose speech bubble says

"Don't panic. But be careful."[36] What Wiles, the microbiologist with a strong commitment to public health communication, asked Morris to help her depict in these frames is what she refers to in an interview as "two attitudes": "I wanted to show these two attitudes. If we had this attitude, this will happen. And if we have this attitude, this will happen. I wanted those two things. I said it would be awesome if we could toggle between the two like some kind of gif."[37]

On the other side of the globe, four days prior to Ardern's announcement in New Zealand, and one day before the WHO declared COVID-19 a global pandemic, Dr. Anthony Fauci, the director of the National Institute of Allergy and Infectious Diseases in the US and the familiar face of the White House Coronavirus Task Force, invoked the epidemic curve not with graphic illustrations, but with words and gestures.[38] "If you look at the curves of outbreaks," he told reporters at the March 10 briefing, "they go big peaks, and then come down. What we need to do is flatten that down. That would have less people infected. That would ultimately have less deaths. You do that by trying to interfere with the natural flow of the outbreak."[39] With the phrase "big peaks," Fauci used a large, upward-curving gesture (figure 4.6), his arm extending above his head. And then with "flatten that down," he used a two-handed gesture (figure 4.7), as if he himself could—and would—do the flattening.

In a March 31 briefing, as the US's death toll from COVID-19 soared

4.6 Dr. Anthony Fauci describing an epidemic curve at a press briefing on March 10, 2020.

4.7 Dr. Anthony Fauci describing flattening an epidemic curve at a press briefing on
March 10, 2020.

and New York City became the new epicenter of the global pandemic,
another member of the White House task force, Dr. Deborah Birx, delivered a detailed presentation with five slides. Figure 4.8 shows Birx
combining gesture and image during the question-and-answer session
to explain the need to flatten the epidemic curve.

By early April, references to the curve were everywhere in the US:
billboards and pledges exhorted the public to "flatten the curve." The
phrase was included in Fauci bobbleheads making a one-hand flattening gesture, which became available for presale. Gestures helped convey the curves, as did metaphors. One language columnist compared
the unmitigated curve to a camel hump and the flattened one to a turtle shell.[40] And later that month, the US Surgeon General Dr. Jerome
Adams tweeted a photo of himself wearing a "flatten the curve" t-shirt.[41]
News stories began to resemble epidemiology manuals, with one
March 19 New York Times article titled "Which Country Has Flattened
the Curve for the Coronavirus?" reprinting no fewer than ninety curves
populated with infection numbers from a wide range of countries.[42] A
similar March 27 article in the French newspaper Le Monde reproduced
178 epidemic curves to show which countries had at that point flattened
the curve—"aplati la courbe"—and which ones had not.[43]

My point in sharing these examples is to remind readers that at the
beginning of the COVID-19 pandemic, the graphic of the two epidemic
curves circulated in a variety of material forms, and that it proliferated

4.8 Dr. Deborah Birx explaining what it means to "flatten the curve" at a March 31 White House press conference. AP photo: Alex Brandon

wildly, often with the needed action stated simply, imperatively: flatten the curve.[44] It is also worth noting the contrast between Ardern's approach and that taken by Drs. Fauci and Birx. In contrast to the United States President, Prime Minister Ardern took time to hold up the images of the epi curve with two possible future scenarios, framing them as a choice that she and her administration—and the country—were facing together, and tying that choice directly to restriction decisions. Fauci and Birx, by contrast, were brought forward to update viewers, which entailed explaining epidemics to the public and to the media. Fauci perhaps moves too quickly through his explanation of how curves work, and what needs to be done, and Birx's discussion became bogged down by multiple charts of the epidemic in progress. Laudatory studies of Ardern's exemplary crisis leadership were quick to follow.[45]

By the end of March 2020, the flatten-the-curve graphic had become, in the words of Harris, "the defining image of the COVID-19 pandemic."[46] It is no exaggeration to say the graphic achieved iconic status. The word "iconic" here comes from Robert Hariman and John Lucaites, who use it to describe publicly circulating photographic images. Hariman and Lucaites define iconic photographs "as those images that are widely recognized and remembered, are thought to represent an important historical event, evoke strong emotional identification or response, and are appropriated across a wide range of media, genres, and topics."[47] The flatten-the-curve graphic or data visualization meets all of these criteria, and yet it is, visually, a far cry from the arresting photographs studied by Hariman and Lucaites, or the moving FSA photographs documenting the widespread suffering during the Great Depression that, according to Finnegan, "helped viewers negotiate emergent anxieties and crises of public life."[48] It does not exude the contagious charisma of a 2006 photograph of Barack Obama, which the street artist Shepard Fairey made into the Hope image, the visual anchor of Barack Obama's campaign and subsequent presidency, as Gries demonstrates in her book-length study of it.[49]

It is rare for a data visualization graphic to achieve iconic status; indeed, in data visualization (dataviz), the word "iconic" might be more likely to name the use of icons to depict information (for example, rocket ships in place of bars on a bar graph), which Edward R. Tufte believes can interfere with good design.[50] Two exceptions can be found in climate visualizations: "the hockey-stick graph" and the Keeling curve, line graphs that show, respectively, the rise in global temperature and the steady rise of CO_2 in the atmosphere.[51] I return to these two graphics later in the chapter, because they support the point I want to adapt from visual rhetoric scholars who focus on photography and digital art. As I mentioned in the introduction to this chapter, like the photographic medium, the medium of data visualization can convey magnitude, and it can carry feeling or evoke emotions, to use Hariman and Lucaites's description. The magnitude shown in these two side-by-side curves, though, differs somewhat from the magnitude that photography is able to mobilize, and that difference helpfully augments the growing vocabulary we have available to talk about magnitude in the context of rhetoric.[52] The flatten-the-curve graphic depicts a distinctly mathematical magnitude, one plotted in a way that makes apparent— emphasizes—the size and rate of growth. Brought before the eyes of the nonexpert public in the midst of a pandemic, quantitative magnitude accrues an eye-widening, feeling-laden weight that conveys information and urgency all at once. Data visualizations can generate felt, qualitative magnitude by means of quantitative magnitude.

Harris's rush to get the graphic in front of the eyes of the public led to a simple design—one that his teenaged daughter apparently mocked.[53] Yet as far as the compositional features of the iconic data visualization go, Harris's version is more than serviceable. Indeed, it meets many of Edward Tufte's criteria, laid out across three books, for "graphical excellence." The labels are clearly placed on the graph itself; the colors of the two epi curves contrast suitably. For the steep gradient Harris opted for red, widely known in graphic arts to convey alarm or warning, and a color that, as Tufte notes, contrasts with every other color.[54] The second curve is presented in a comparatively muted blue.[55] The most overt purpose of the graphic is, again, to encourage comparison between the two curves—encouraging the eye to compare is, for Tufte, a hallmark of graphical excellence—so that viewers can see for themselves the wisdom in striving for the second curve as the more gentle, less chaotic future. But when the graphic was imported into climate change, it quickly became clear that the iconic curves carried from the pandemic much more than a bare message of comparison.

Climate Curves

During the first two months of the pandemic, the epidemic curve and its message (to flatten it) were redirected toward another pressing crisis: that of a rapidly warming climate. The appropriation itself happened with astonishing speed. On March 31, 2020, the same day as Birx's five-slide briefing at the White House, David J. Hayes, then director of The State Energy and Environmental Impact Center at New York University, published a piece featuring a near replica of the image of the by-then-familiar epidemic curves (see figure 4.9).[56]

In this graphic, the vertical axis depicts not a rising rate of infections, but the rise of global temperature—first without mitigation, and then with it. The horizontal line that shows the limit of treatment capacity in epidemic graphs here depicts a two-degree limit on temperature rise, the amount of rise during the current epoch that climate scientists have determined will wreak havoc on this planet's life forms.[57]

Here is the opening line of Hayes's reflection on the graphic:

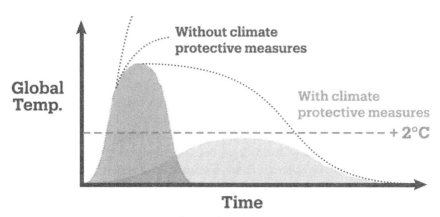

4.9 David J. Hayes, "Flatten the (Climate) Curve." Design credit: David J. Hayes and Stephen Read.

"COVID-19 is teaching us an important lesson about 'flattening the curve.'" He continues, explicitly adapting that lesson to climate change:

> This flatten-the-curve teaching also applies to the climate crisis. The timeline differs and, thankfully, the strategies for attacking climate are far more economically palatable than those for the coronavirus. But the message is the same: If we take action now to decarbonize our economy, we have a fighting chance to keep the global temperature rise in check, avoiding the most catastrophic consequences of climate change. Indeed, because of the long life of the climate-damaging greenhouse gases that we are adding to the atmosphere, if we miss our chance to shave off peak emissions, we may be living "above the line" with a climate catastrophe for a very long time.[58]

Noteworthy in this excerpt is how Hayes frames the phrase "flatten the curve," first as a lesson, and then as "teaching." At least part of that teaching is happening through sight. The curves and what they portend relative to each other were by this point familiar to a broad cross-section of the public, as evidenced by their viral circulation. By repurposing the pandemic's epidemic curves in a climate-oriented data visualization, Hayes effectively translated the climate crisis into terms the public was just at that point beginning to grasp, thereby tapping the new pandemic-related knowledge being sought and gathered by the public. This merger of the two crises' curves offers yet another instance of affective images sticking together through association.[59] But that stickiness holds a number of implications for rhetoric. As Hayes's brief discussion makes clear, the visualization carries feeling as much as fact: the limit line of two degrees conveys the number we ought to do all we can to avoid. The limit line, in both graphics, depicts urgency, the point of overwhelm, what Laurie Gries calls "consequentiality."[60] Hayes and his colleagues also modified the graphic by adding three dotted lines like the ones used in IPCC reports to indicate future projections based on different computer-modeled scenarios.[61] The first two of these lines are steep, and they resemble the famed hockey-stick curve pictured in figure 4.12. The reddish dotted line that evens out, though, acknowledges the imperfect comparison: the speed of viral burnout, usually represented by a steep downward slope, does not apply as easily in the climate crisis. As Hayes puts it in his explanation, "if we miss our chance to shave off peak emissions, we may be living "above the line" with a climate catastrophe for a very long time." The x-axis of Hayes's repurposed graphic does not give a specified timeframe, but given the speed of viral

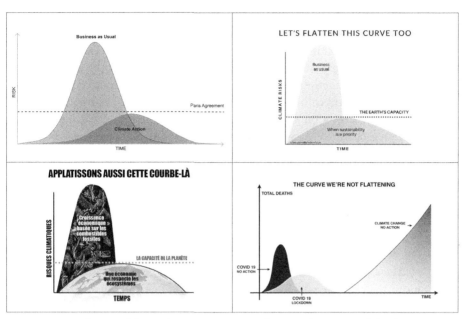

4.10 A selection of flatten-the-curve images circulating in the context of the climate crisis. Credits, in order, beginning with top left: (a) Joel Makower / GreenBiz Group (greenbiz.com); (b) Sustainable Fashion Forum; (c) La Planète s'Invite au Parlement; (d) Nicholas Gruen, Lateral Economics.

spread relative to atmospheric warming, it's likely that his x-axis spans decades or centuries rather than weeks or months.

Hayes was not the only one to appropriate the pandemic's iconic graphic and its slogan for the climate crisis in the early months of the COVID-19 pandemic.[62] Figure 4.10 shows four such images that circulated online.

Note that the climate curve figure 4.10d, with no measure specified by a dotted line, slopes right off the chart into a seemingly inaccessible, even unimaginable future. The other depictions compare mitigation to no mitigation, with the French-language one cleverly designing the first curve to resemble a mountain of coal and the second curve the planet as viewed from space. Directly reproducing the two epi curves, the point in "The Curve We're Not Flattening," drawn by the economist Nicholas Gruen, is to compare crises, to suggest that the climate crisis's magnitude will outstrip the pandemic's, which in this graph seems tiny by comparison. This visualization uses a mix of comparative scale and chart area to stretch imaginations and tip into sheer feeling. Its implication: "If you think this pandemic is bad, just wait."

From Bending to Flattening

The idea of flattening a climate curve did not remain confined to line graphs. It also appeared as cover art. The cover of the May 23, 2020, issue of *The Economist* shows an arcing cloud of smoke, emitted from a smokestack, broken by a white fist clenching part of it (figure 4.11).[63] The image draws on one of climate change's commonplace images (a smokestack) while also invoking, mainly by word this time, the iconic flatten-the-curve graphic.[64]

This issue of the magazine features a clutch of articles presenting the pandemic as a "unique chance" to remake economies devastated by the virus in a way that promotes green-energy infrastructural development and simultaneously curbs greenhouse gas emissions, thereby, in the words of one the titles in the suite of pieces, "Flattening the Other Curve."[65]

The special issue of *The Economist* leans on the word "flatten," even more than on the visual curves, to make the link to climate change, thereby highlighting one other tweak that the pandemic brought to the climate crisis's existing treatment of curves: the use of the verb "flatten" in relation to a rising curve.[66] The verb "flatten" replaces the verb "bend," which epidemiologists and climate scientist alike had used to refer to key line graphs prior to and early on in the pandemic.[67] The UN Secretary General's statement accompanying the release of the 2018 IPCC report on 1.5°C offers a representative instance: "The coming period is critical. We must meet the Paris commitments to bend the emissions curve by 2020."[68] And as the COVID-19 epidemic became a pandemic, the verb "bend" was still being used, as when one Johns Hopkins University epidemiologist told STAT, the health and media outlet: "it's the top priority right now that we bend the curve."[69]

Bending the Curve is the title of a report and 2019 book, both created by V. Ramanathan, a climate scientist at the University of California, San Diego.[70] Ramanathan's preface offers this reflection on that title, particularly the word "bending," and what exactly is getting bent:

> The title of this book refers to the resulting rise in global temperature, represented as an ever-steepening curve over time. Bending that upward curve to decrease the unsustainable trajectory of an increasing global temperature requires a significant focus on reducing the release of emissions of carbon dioxide and four short-lived climate pollutants into the air. Without mitigation, the warming will reach dangerous levels before 2050 and we will be transitioning from climate change to climate disruption. The timeline for bending the

4.11 *The Economist* cover "Seize the Moment: The Chance to Flatten the Climate Curve."
© The Economist Group Limited, London, May 23, 2020.

warming curve is aggressive. Mitigation actions have already begun
in many cities, states, and nations. It must proceed at a rapid pace
such that emissions of all climate warming pollutants will be reduced
by 50% to 80% by 2050, followed by ongoing carbon neutrality before
2100. We must also be prepared to extract as much as 500 billion to
a trillion tons of carbon dioxide from the air during this century.
Bending the curve of climate change has emerged as the challenge of
our time.[71]

One challenge with "bending the curve" as a slogan is that the climate crisis has generated so many different curves. Even after this clear indication that the curve in question is the rising global temperature, the discussion of the book's hoped-for readership ("Who Is This Book For?") six pages later ends by noting that we can all "work to bend the emissions curve."[72] In some ways, the slip from warming curve to emissions curve is understandable, even to the point where perhaps it's not a slip at all. While not the same thing, the warming curve and the emissions curve are surely correlated, as the ten solutions outlined in Ramanathan's chapter 4 indicate: "This book is organized around a set of ten solutions, designed to bend the curve—to reverse the trend of increasing greenhouse gas emissions and keep the planet below dangerous levels of warming."[73] *Bending the Curve*'s figures include a number of different curves—the Keeling curve, a global land-ocean temperature curve, probability curves for temperature rise based on severity of mitigation, and the list goes on.[74]

Another challenge with the word "bending" as compared with "flattening" is that the gerund "bending" does not specify which way the curve needs to be bent. What both words seek, of course, is an inflection point, the term used in geometry to indicate the point at which curvature changes.[75] Of course the term "inflection" can also carry a figurative, moral valence.[76] But the point of bending, really, is that the line graph charting the rise of CO_2 over the long course of the current geographic epoch (the Holocene) and the one charting global temperature rise during the same period have yielded a straight line of compound, accelerated growth. The definition of the verb "to bend" marked in the *OED* as "now the main sense" is one that inflects something already straight, making it into an "arched or angular shape."[77]

"Flattening the curve" acknowledges the steep rise and gives a specific direction—downward from above. It entails deliberate suppression. In fact, the action of flattening the curve, as Fauci's gesture suggests (see figure 4.8 above), can be produced through effort, comes from above, is deliberate, and is agentive, which is to say, it requires a concerted exertion of energy. As a verb, "flatten" appears in all sorts of contexts, most commonly perhaps in economic ones, where markets or prices can be flattened or depressed by certain influences. But the word taps a number of sensory modes: winds, paints, and musical notes can all flatten or be flattened. "Flatten," like "bend," can be transitive or intransitive, which means the verb can be used with or without an object. In this case it is, of course, transitive—the curve is what gets (or needs to be) flattened. Many of the current uses outlined in the *OED* are subtended by a meaning now considered obsolete: "to deprive of energy or 'fire.'"[78]

The energy or effort embedded in, required by, the imperative verb in the slogan "flatten the curve" may well be key to the phrase's brief rise to the status of rallying cry.[79] Indeed, to highlight the difference, headline writers for stories covering the wild successes in curve-flattening in nations like New Zealand and Iceland began to reach for more energetic verbs—squash, vanquish, and (a favorite) crush.[80] The imperative verb "flatten," used transitively with "the curve" as its object, became enough to conjure on its own the data visualization discussed at length in this chapter. The phrase, in effect, began to carry with it the image of the graphic as well as the complex of feelings the image gathered as it circulated, many of them related to impending danger and an urgent need for action.

That the word "flatten" itself has begun to conjure multiple magnitudes is all the more evident in the phrase's leap to climate curves. In "Will We Be Able to Flatten the Climate Curve?" the journalist Nick Clark, the environmental editor at *Al Jazeera English*, focuses not on the curve of rising temperature, but on the other *other* curve, or what is known as the Keeling curve, which documents the rising level of CO_2 in the atmosphere. These two curves—the global temperature curve and the Keeling curve—remain, as I mentioned above, the most familiar curves of climate change data. The hockey-stick graph, reproduced in the 2001 IPCC report and pictured below (figure 4.12), gets its

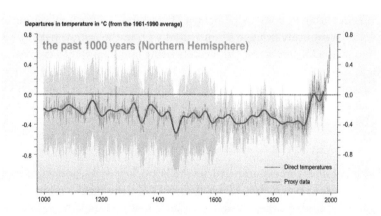

Departures in temperature in °C (from the 1961-1990 average)

the past 1000 years (Northern Hemisphere)

——— Direct temperatures

——— Proxy data

4.12 The hockey-stick graph. From R.T. Watson and the Core Writing Team of the IPCC, et al., *Climate Change 2001: Synthesis Report*, figure 3. © IPCC 2001. Published by Cambridge University Press.

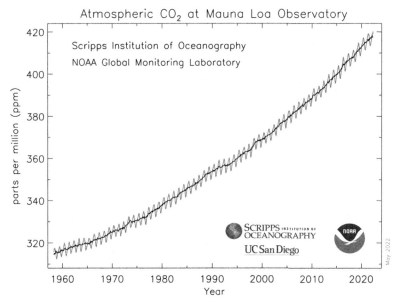

4.13 The Keeling curve. Credit: NOAA and Scripps Institution of Oceanography.

colloquial name from the shape the curve takes in the last hundred or so years: the curve rises so sharply that it starts to resemble a hockey stick lying on its shaft, its blade curving away and up.[81]

The other curve often referred to as iconic, especially among climate scientists, is the Keeling curve, a line graph with a distinctive upward slope (see figure 4.13—the regular jags are the seasonal fluctuations, considered analogous to planetary "breathing"). The Keeling curve plots the monthly average of measured CO_2 in the atmosphere at the Mauna Loa Observatory. The steadily rising numbers yield a curve that bears some resemblance to the long and rising slope of the Hawaiian volcano where the observatory is situated (Mauna Loa is Hawaiian for "Long Mountain").

Picturing Together

What, besides camel humps, turtle shells, hockey sticks, and volcanoes, do all these curves picture? First, the hockey-stick graphic and the Keeling curve both picture unprecedented growth. Here is how Michael Mann, the lead author of the pair of landmark studies that presented the temperature graph, describes the work of those articles and their graphic (the two slide into each other syntactically and grammatically): "The original hockey stick articles presented this temperature curve going back a thousand years that demonstrated really how unprecedented modern warming is."[82] The synthesis of the 2001 IPCC report reprints the hockey-stick graph as broader context for a graph showing direct temperature measurements from 1860 to 2000. The figure includes a long caption beginning with these words in bold: "The Earth's surface temperature has increased by about 0.6°C over the record of direct temperature measurements 1860–2000 (top panel)—a rise that is unprecedented."[83] An article published in *Science* extending Mann's studies to the entire Holocene (11,300 years) titles the abstract with the phrase "Exceptional Now."[84] So too with the Keeling curve, which charts the concentration of carbon dioxide in the atmosphere. Each spring, when CO_2 is at its annual peak, the Keeling curve passes a new milestone. The year 2020 was no exception, in spite of the temporary production slowdowns and closures during the COVID-19 pandemic.[85] That spring, the new record measurement was set at 417.2 ppm, up from the previous year's 415 ppm, which, you will recall, is the measurement inscribed at the bottom of the plaque memorializing Okjökull, the Icelandic glacier discussed in chapter 2.

But to read the current levels—1.1°C, 417 parts per million—is one thing. To see them graphed in relation to previous levels is quite another.[86] Figure 4.14 shows the Keeling curve not in its familiar annual rise since Keeling began measuring atmospheric CO_2 in 1958, but on a thousand-year timescale, like Mann's, the pre-Keeling measurements plotted using proxy data: those atmospheric gases trapped in ice cores that I discussed in chapters 1 and 2. This graph, too, starts to take on the recognizable hockey-stick form; the data from the past century seem to lose their nuance when presented at this timescale. This curve features prominently in *An Inconvenient Truth*, the 2006 documentary film about Al Gore's quest to educate Americans on the threat of climate change.[87] To materially emphasize the point of CO_2's extraordinary rise, Al Gore famously had a cherry picker carry him to the top of the now vertical line.

These line graphs, like the flatten-the-curve graph, picture unprecedented growth, yes. But the unprecedented nature of the growth is probably not why audiences react audibly when they are shown (by Mann or by Gore) either hockey-stick-shaped curve.[88] That collective gasp more likely owes to compound growth—to mathematical magnitude.

As I discussed in this book's introduction, magnitude is a weighted rhetorical feature, one that does not just direct attention, but holds it. Mathematical magnitude, specifically, works rhetorically by collapsing the distinction between calling attention to something's enormous importance and presenting enormity itself. So far, though, scholars in

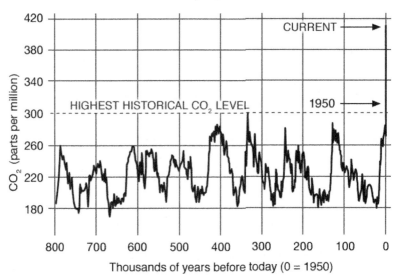

4.14 Historical CO_2 levels. Data: NOAA.

rhetoric (myself included) have attended to magnitude as a quality of size—Christa Olson draws attention to the concept's dimensions of grandeur, Jenny Rice to its weight, Stephanie Larson to its accumulation.[89] Farrell, whose work ignited the field's interest in the concept, charts a grammar of magnitude, finds its work in sets of incongruous conceptual pairs (e.g., more/less, gigantic/tiny, enlarge/diminish, appreciate/depreciate).[90] Taken together, Larson's and Rice's treatments of magnitude, and several of Farrell's pairs, have a quantitative valence. But the flatten-the-curve graphs foreground most expressly the mathematical conception of magnitude. In math, the phrase "orders of magnitude" names sets of numbers as they compound, typically by powers of ten.[91] When the number of confirmed COVID-19 infections passed 1,000,000 in the United States on April 28, 2020, for example, the cases had increased by six orders of magnitude from the first case. That they did so in about three months is, to many of us nonexperts, mind-boggling, unfathomable. Even less fathomable are the two weeks in July 2020 when the number of known infections in the US jumped from three to four million. Yet to epidemiologists, economists, and mathematicians, such accelerated spread is less surprising; this is simply how exponential growth works. For those of us studying rhetoric, magnitude's conceptual latitude—the way it ranges between a specific mathematical concept (an order of magnitude) and conferral of value, felt weight—is a feature of its potency worth dwelling on, for it names and distinguishes the kind of magnitude data visualizations can convey.

Quantitative Eloquence, Exponential Difficulty

The idea of quantitative eloquence is familiar to scholars in technical communication and design rhetoric thanks in no small part to the work of Edward R. Tufte, who devoted the better part of his career and three stunning books to documenting the "tremendous communicative power of graphics."[92] Tufte lays important groundwork for understanding mathematical magnitude's rhetorical side. His third book opens with two observations and three questions, all elegantly stitched together:[93] "Our thinking is filled with assessments of quantity, an approximate or exact sense of number, amount, size, scale. In scientific work, both order-of-magnitude reasoning and precise measurements are pervasive. How are such quantities represented in visual expressions of ideas, experience, evidence? How are moving images, photographs, diagrams, maps, and charts to be scaled and labeled? And what makes images quantitatively eloquent?"[94]

When Tufte poses the question "How are such quantities represented in visual expressions of ideas, experience, evidence?" he links reasoning modes of "assessments of quantity" and "order-of-magnitude reasoning" with their visual expression. From there, the passage builds, efficiently, to a concept of quantitative eloquence, a concept that has by the end of the passage absorbed the idea of visuality so thoroughly that it need not mention it. In his first book, *The Visual Display of Quantitative Information*, Tufte, in a characteristic celebration of efficiency, observes that "only a picture can carry such a volume of data in such a small space."[95] In a small space, graphs—especially elongated ones like the hockey-stick graph—can compress vast swaths of time, allowing viewers to glimpse astonishing rates of change and leaps in relatively short time periods. This is why Michael Mann's audiences so often gasp when he shows them the hockey-stick graph: the great power of data visualization is its ability to bring before the eyes, to help data (Latin: "the givens") move, to make sense at a glance.[96]

Humans need that assist. But even with it, their ability to contemplate a single, vast object is limited. Aristotle parlayed this point into advice for aspiring playwrights when in the *Poetics* he wrote that humans are unable to picture an animal a thousand miles long: "Besides, a beautiful object, whether an animal or anything else with a structure of parts, should have not only its parts ordered but also an appropriate magnitude: beauty consists in magnitude and order, which is why there could not be a beautiful animal which was either minuscule (as contemplation of it, occurring at an almost imperceptible moment, has no distinctness) or gigantic (as contemplation of it has no cohesion, but those who contemplate it lose a sense of unity and wholeness), say an animal a thousand miles long."[97]

Aristotle here identifies what is, at base, a persistent perceptual difficulty of wrapping our minds around objects of extraordinary size. Peters helpfully puts numbers to the limits in his meditation on the infinitesimally small: "Human sensory experience that is not technologically supplemented takes place generally within the range of about 10^{-3} meters (grains of sand) and 10^4 meters (views of the horizon or of thunderclouds)."[98] These are physical limitations beyond which perception requires assistance. These limitations, in other words, require accommodation, from the Latin *commodus*, a fitting, with the prefix *ac*, one thing to another. The faculty of *phantasia*, what the Greeks called imagination, as I have argued elsewhere, often steps in to help with that fitting. Imagination can help compensate for that which is too large or too small to be sensed or grasped.[99] Viewed from this rhetorical perspective, the flatten-the-curve graphic works as an aid to the imagination, plotting numbers enormous and less so, urging action to favor the less drastic growth. And all this rhetorical work is compressed into a single graphic that can be taken in at once. My point here is not to say "hey! Graphics present information in compact and powerful ways!" That point has been made by the likes of Tufte and a long line of others.[100] The point is to dwell on the extra-informational features that this short-lived crisis-to-crisis adaptation reveals—to dwell, that is, on the way different kinds of magnitude layer together and even collide.

But the scholarship on rhetorical magnitude, following Farrell following Aristotle, has so far concerned mainly visual, material *objects*, aesthetic objects such as photographs, a column, an archive, an extremely long list. The flatten-the-curve graphic, a visual object too, no doubt, still works in its own way. Whereas the American landscape images examined by Olson and the large and looming Roman column considered by Jonathan Mark Balzotti and Richard Benjamin Crosby evoke grandeur of size, and the conspiracy theorists whose archival evidence (in Rice's account) exerts sheer weight, Finnegan and Larson's studies lean more toward the heaviness of feeling conjured by the evidence of magnitude. Finnegan's archive, for example—comment cards filled out by viewers after viewing an FSA photography exhibit—offers insight into the specific ways the photographs helped museum visitors negotiate the Great Depression's magnitude; but of course the photographs first had to make that magnitude apparent, to bring it before the eyes, to conjure its enormity. Magnitude, in that case, led some of the viewers to plead for government action.[101] Larson's analysis of the rhetorical effects of lists draws a line from the #MeToo movement's "massive and daunting list"—from the enormity of the accumulation of testimony—to a visceral understanding of rape culture. In this way, magnitude can, as Larson puts it, "jolt audiences into recognizing their own bodily vulnerability and, ultimately, inspire change."[102] And in the words of Balzotti and Crosby, "magnitude works not just to engage viewers but also to overwhelm them."[103] Magnitude, in these ways, carries with it the possibility of transformation—though importantly, not the certainty of it.

Indeed, it is helpful to linger on the magnitude-focused analysis that Balzotti and Crosby offer of the thirty-meter column erected in Alexandria at the beginning of the third century BCE to commemorate Rome's victory there, because the giant column is somewhat easy to imagine, looming as it did far above existing Egyptian monuments, departing from them in both style and size, an icon to Rome's dominance. Diocletian's column effectively commemorates Rome's imperial victory by replicating that victory's significance: the column, as Balzotti and Crosby point out, overtakes the landscape such that, as they put it, "existing subjects are now understood in relation to the new, ostensibly dominant subject."[104] The column becomes, in Balzotti and Crosby's analysis, a material marker of disruption, a reminder of the power of *megethos* (the Greek word for magnitude discussed in this book's introduction) to disturb, to reorder, and to constantly assert itself, loudly and largely.

The column and its long shadow create magnitude by overwhelming viewers and generally disturbing and remaking the world. So too with the flatten-the-curve graphic, only the graphic disturbs—looms large—by plotting overwhelm itself (overwhelm of hospital systems, Earth systems) and predicting the overturning of the known world if action is not taken. The graph's magnitude, in this way, effectively compounds.

The magnitude pictured by the flatten-the-curve graph—and by the hockey-stick graph too—is a rate of growth, exponential growth at that. Conveying such conditions calls for what Tufte labels "quantitative eloquence." Far from lying outside rhetoric's art, in a corner where Aristotle shoves legal contracts and testimony under torture, the tools for quantitative eloquence reside—if perhaps latently—in rhetorical concepts such as amplification (Latin *amplificatio*: a widening, enlarging, extending, increasing) and perhaps even emphasis, which derives from the verb *emphanaien* (to display, to indicate, but also, when used in middle or passive voice and with future forms, to "become visible").[105] Indeed, quantitative eloquence results from effectively displaying quantitative increases (or decreases) that are of wider concern.

Take growth rate. In an unmitigated epidemic, cases can increase at an exponential rate (up to the population limit, at which point decline will be just as precipitous), a phenomenon that is itself difficult for most humans to grasp. What begins as one or two known infection cases in a region can multiply throughout a population. With COVID-19 in Manhattan, for example, painstaking news reports on the movements of a single infected midtown attorney in March 2020 became, within three weeks, stories about sheer numbers, about the disease churning through all boroughs of New York City, its hospitals overwhelmed with COVID-19 patients.[106] The number of infections in the city was great, and that number increased at an astonishing daily rate. And even after government and public health officials in New York worked to decrease the rate of infections, leader after leader in the US and elsewhere continued to downplay or even dismiss the possibility of such growth in their areas.[107] As the risk researchers Howard Kunreuther and Paul Slovic put it in a *Politico* article, "the deceptively mild and seemingly faraway beginnings of the current pandemic led health officials and governments to squander many opportunities for early intervention."[108] It's as if Aristotle's miles-long animal lined up for a race at light speed: the magnitude of size and time simultaneously are too much for nonexpert human faculties to handle.[109]

Quantitative magnitude, then, accretes into the kind of magnitude rhetoricians have been concerned with of late. As the mathematical implications begin to "sink in" (Larson's phrase), another kind of magnitude, one laden with feeling, takes hold. As Heather Houser puts it in a slightly different context: "One epistemological mode— quantification—can swerve to produce another—emotion."[110] A conception of mathematical magnitude elucidates how Houser's epistemological "swerve" happens rhetorically, through a complex layering of quantitative magnitude with qualitative, experiential magnitude such that the two can begin to work in concert, each amplifying the other.

To see the steep slope of the epidemic curve when a virus is tearing through a population is to encounter a picture of exponential growth, a mathematical concept of compounding accumulation—things that grow in increasingly large increments, by multiplication rather than addition (the latter is linear growth). Exponential growth is famously difficult for human minds to grasp in the abstract—it is a concept that lies beyond most humans' intuition. Since the 1970s psychologists and economists have proven the concept's difficulty so consistently that they identified what is now known as "exponential growth bias," which names "the systematic tendency to underestimate compound growth processes."[111]

Line graphs and their counterparts (scatterplots and the like) have been found to be no more successful than straight numbers in helping people comprehend exponential growth.[112] But human eyes (and minds) do better with comparison. So the flatten-the-curve graph's initial effectiveness owes in part to the presentation of both curves in the same graph, or in the case of Ardern's address, one after the other. The graph with Harris's hospital capacity line allows viewers to glimpse the implications, if not the mathematical basis, of such exponential growth and of the astonishing and uncontrolled rise in infections, especially when brought before the eyes next to the relatively gently upward-sloping, and ultimately much flatter, curve of mitigated spread. That distinction, according to Harris (who, recall, added the system capacity line to the flatten-the-curve graphic), can—ought to—carry a sense of urgency. In an opinion piece published in the *Philadelphia Inquirer*, Harris translated the curves, and the epidemic scenarios they depict, into feelings and actions:

> The high peak is selfishness, disregard for our neighbors, and fear of the foreigner. It means fighting for precious and dwindling supplies at the market and begging for lifesaving services at the hospital. It is chaos.
>
> The low curve describes a challenge contained—triumph over a deadly threat through feats of collective action small and large. It depicts coming together as a community to preserve precious capacities so we don't have to decide who lives and who dies.
>
> The low curve is love and caring. It's people buying groceries for their elderly neighbors too scared to leave their house. It's families isolated behind closed doors singing out open windows spreading a joyous sound that travels where no virus can go.[113]

What's noteworthy here is how Harris characterizes actions by dispositions of people toward each other—by, really, public feelings. And those feelings can only be described by imagining the scenarios depicted in the flatten-the-curve graphic. The two curves, that is, project more than numbers. They go, as the title of Harris's piece indicates, "way beyond science." As such, as the image of the curves circulated in March 2020, they accrued public feeling. And they did so by helping people glimpse the magnitude of taking action—swift, caring action—by teaching them to see (and then imagine) the magnitude of the consequences of *not* taking action. That a visualization of two epi curves can—and perhaps temporarily did, in California and the northeastern region of the US—press state and local leaders to act owes at least in part to the imaginative work required to play out the scenarios and to imbue them with feeling.

Just as Diocletian's column loomed over Alexandria, altering the landscape and creating new relations of understanding (as Balzotti and Crosby argue), so too did the flatten-the-curve graph—for a short time—change the data visualization landscape, bringing forth new ways to present (make present) the climate crisis and its key graphics. On March 10, 2020, the climate economist Gernot Wagner used the emerging epidemic to clarify and amplify the climate crisis, pointing expressly to growth rate when he tweeted that COVID-19 "is climate change on warp speed." The tweet continues:

> It's *not* about levels, it's about growth rates.
>   Yes, 1°C is bad. But it's nothing compared to +2, 3, or 4 expected by 2100.
>   Well, a dozen Covid cases in NYC is bad. But it's nothing compared to 1,000.[114]

It is important to note that temperature rises in a linear fashion (through addition), though because of its rate of accumulation, the effects of that linear rise—such as increase in number of days in a year that threshold temperatures are exceeded—are, according to the IPCC, "approximately exponential."[115] Carbon dioxide in the atmosphere is growing exponentially.[116] And exponential decay, which is now being used to characterize ice melt, names another almost unfathomable mathematical magnitude.[117]

In other words, the steep curvature of temperature rise and the acceleration that curvature depicts guides the conversation comparing climate change to the pandemic. An editorial in the special smokestack issue of *The Economist* mentioned above begins like this: "Following the pandemic is like watching the climate crisis with your finger jammed on the fast-forward button."[118] Another popular press piece titled "Seeing the COVID-19 Crisis Is Like Watching a Time Lapse of Climate Change. Will the Right Lessons Be Learned?" was co-written by the moral philosopher Lawrence Torcello and Michael E. Mann, the climate scientist of hockey-stick fame.[119] Climate change data began to be understood and framed in relation to the new, more dominant subject (the pandemic). One crucial lesson that climate scientists and activists sought to carry with the flatten-the-curve graph from the pandemic into the climate crisis is the urgency of *early* action. Mann expounds on this lesson in his 2021 book, in a paragraph that begins, "The coronavirus outbreak also taught some important lessons about the cost of delay."[120] With exponential growth, unchecked numbers can spiral out of control, and they can do so at what is, for most, unexpected speed. Even as late as July 20, 2020, leaders such as Idaho's governor were dismissing small numbers of infections that were, effectively, doubling.[121] But the numbers always start small; the conditions seem mild to most everyone— policy makers, citizens. And then, suddenly, the same people, sanguine about the numbers, find themselves in the middle of a full-blown crisis. As it plays out, this sort of magnitude—mathematical magnitude—says not only "Look at this! This is important!" (Farrell's formulation), but: "This is *dire*."

People in countries that were early epicenters watched with something like dread, if not horror, while leaders in other countries delayed action. Early intervention was rewarded with a drop in infections and low case rates. The early, hard strategy is where smaller countries like New Zealand and Iceland succeeded, and where places like the US— late, dithering, skeptical—failed.[122] A reporter on location in Tuscany, a region south of the early and deadly epicenter of the epidemic, relayed what it felt like watching the pandemic's arrival in the US. "Friends of mine told me they feel the reports from the U.S. are like reading science fiction or watching a disaster movie, you know."[123] The reporter layered the intensity of experience and memory onto the imagined realities conjured by reports from the US: "The horrible images from March of army trucks transporting coffins that had been piling up outside of cemeteries in Lombardy are very much alive in the Italian consciousness."[124] For people living in former epicenters, these images are attached to the pandemic, their experience becomes embedded in epidemic curves and how fast they can rise. If most Italians, like most humans, don't quite grasp the concept of exponential growth, they nevertheless had at that moment a deep understanding of what the steep curve portends. In this way, different kinds of magnitude, all of which feed each other and create a sense of urgency, combine for an intensely felt rhetorical relation to numbers.

In this way, the pandemic's lesson of exponential growth is clear, and it aligns with the message delivered by New Zealand's leader as she showed the flatten-the-curve graphic: go hard, go early. The need for early action has been the plea of climate scientists all along and has, for the most part, gone unheeded. The 2019 UN emissions gap report makes clear that the longer action is delayed, the steeper the emissions cuts will need to be. That fact owes, of course, to exponential growth rates.[125] The director's foreword to that report laments the lack of action to this point: "Our collective failure to act strongly and early means that we must now implement deep and urgent cuts." And "We have to learn from our procrastination. Any further delay brings the need for larger, more expensive and unlikely cuts."[126] Learning from procrastination can itself be a challenge, especially in what, to many, seems like a slow-moving crisis. But the much faster-moving pandemic offers vivid, present, and *felt* lessons of compound growth rates.

Conclusion: Crisis-to-Crisis Rhetoric, Felt Magnitudes

The premise of this book is that the climate crisis is changing rhetoric, making—and making available—new means of intensification. Intensification, as I discussed in this book's introduction, is, after all, the primary work of rhetoric. Magnitude is a—if not *the*—lead concept in that intensifying work.

The COVID-19 global pandemic careened in with a deadly illustration of that very premise, itself mingling with and providing new lessons for climate communicators, lessons this chapter has drawn out. What persists in my account is the *sense* of one crisis temporarily carried into another. As the climate activist Bill McKibben put it in the context of the pandemic: "So now we *have a sense* of what it's like: a full-on global scale crisis, one that disrupts everything."[127] The flatten-the-curve graphic has come to carry the very sense McKibben invokes. The early rhetorical life of the flatten-the-curve graphic in the early weeks of the COVID-19 pandemic in 2020, its development in contexts of teaching and planning, its wide proliferation, and its quick leap from one unprecedented crisis to another, together reveal borrowed magnitudes—mathematical magnitudes that interleave with and mutually intensify each other: felt magnitudes.

Of course, reading this chapter months and even years after the onset—and, one can hope, after the end—of the COVID-19 pandemic raises the question: why didn't the flatten-the-curve campaign work better in places like the United States? The answer may well lie in the prevalence of other kinds of magnitude. As Olson shows, what she calls "American magnitude" has, especially at the pandemic's outset, quite the stronghold in this country. When that magnitude prevails, freedom becomes a tenacious feeling violated by mitigation policies and their supporting forms of information.[128] A leader whose power rests on this kind of resolute American greatness would ultimately not be able to reconcile the two kinds of magnitude.[129] Irreconcilable magnitudes lead to, even perhaps necessitate, denial.[130]

Magnitudes, by their nature, loom large, as this chapter has highlighted. In examining the brief merger of the epi curve with climate curves, and the entry of the slogan "flatten the curve" into public-facing writing about climate change, this chapter has made clear the difficulty and potency of mathematical magnitude. Following mathematical magnitude in its graphical and sloganized form shows a vivid instance of cross-crisis communication at work, one that bears on this book's concern with how rhetorical intensification operates in the absence of precedent.

In this chapter, the heavy and all-consuming experience of another unprecedented situation (the global pandemic) can be seen providing new instruction and newly salient resources for the intensification of sobering, felt lessons on exponential growth and the need to anticipate it by intervening early on. Some people still may not fully grasp exponential growth, and others may turn away from predictions and toward other forms of magnitude, but the point remains that visuals can carry feeling from crisis to crisis. What this brief crisis-to-crisis borrowing of the epi curve graphic shows from a rhetorical perspective is how magnitude, as a prime means of intensification, can be borrowed, its intensities redirected, mobilized anew. Chapter 5 moves into the realm of public climate art, but the pandemic is not left behind. For really, at this point, how can it be?

CHAPTER 5

Presence and Placement in Maya Lin's *Ghost Forest*

"Our trees," proclaims the arboretum web page of New York's Madison Square Park, "give our park life."[1] And yet for twenty-seven weeks in 2021, the park hosted a public art installation composed of forty-nine trees, selected for being very near to death. All the trees in the installation, leafless Atlantic white cedars, had already been tagged for removal from the Pine Barrens in New Jersey, about 90 miles down I-95, to allow for forest regeneration.[2] They were transported late at night to the urban park and secured in place by the same buried hardware and fill used for telephone poles, arranged into a grove on the park's central oval lawn.

The resulting installation, titled *Ghost Forest* (see figure 5.1), was created by Maya Lin, best known to rhetoric scholars (and indeed to a broader public) as the artist who designed the Vietnam Veterans Memorial in Washington, DC, and the Civil Rights Memorial in Montgomery, Alabama. Lin has devoted the last two decades—but really, she told a small gathering of press members and conservancy staff, her whole life—to environmental issues.[3] Madison Square Park Conservancy's chief curator, Brooke Kamin Rapaport, finds "intense power in an artist who uses materials directly from nature to create a work that defines a cataclysmic crisis of our time, the devastation of climate change."[4] That intense power—how it works rhetorically in *Ghost Forest*—will occupy this final chapter. For *Ghost Forest* exemplifies how art, and especially public art, can redirect rhetoric about climate change, can intensify without polarizing.

So far, this book has examined how ceremonial rhetoric, generational testimony, and another, faster-moving, crisis all in different ways reduce the spatial and temporal distance of the climate crisis, thereby heightening its magnitude and creating new forms of witnessing. As Robert Cox

5.1 Maya Lin, *Ghost Forest*, 2021. Forty-nine Atlantic white cedar trees. Collection of the artist, courtesy Pace Gallery. © Maya Lin. Photograph by Andy Romer / Madison Square Park Conservancy. The exhibition was organized by Madison Square Park Conservancy, New York, and was on view from May 10 through November 14, 2021.

and Phaedra Pezzullo note in their environmental communication textbook, "the lack of visual evidence of climate change has been a problem for scientists in educating the public of the problem."[5] *Ghost Forest* moves past or through visual evidence to material and even felt evidence, and it also formulates the problem as no longer that of scientists. To *feel* evidence is to begin to grasp the enormity of the implications—the magnitude of what is happening. Felt evidence weighs and sticks, as in a dense atmosphere where breathing becomes difficult. Felt evidence differs from a calculated, deliberative, rational process of "weighing the evidence," even as it draws on such evidential knowledge, mixing it with other visceral and ethical forms of knowledge—especially Indigenous knowledge.

Public art, because it uses material, sensuous techniques of placement and presence that can hold together, mix, and circulate multiple knowledges and temporalities, can provide felt evidence. *Ghost Forest* is but one of hundreds of public art works that seek to provide climate

information and provoke climate action. Still, this temporary installation presents a particularly poignant example of how public art—
emphasis on public—can overcome the temporal and spatial challenges posed by the climate crisis. And so this chapter contributes to
the emerging conversation in rhetoric about climate art, considering
specific means by which public art makes accessible worlds otherwise
inaccessible. It does so in a way that imposes those worlds physically
and sensuously, presenting (making present) cascading losses—of
tribal lands and peoples, of biodiversity, of coastline—as intimately
related by holding those losses together in one place. In exhibiting
one instance—or a stand of instances—of environmental destruction
caused by a too-quickly warming planet, *Ghost Forest* provides felt evidence by means of material, sensuous presence—first through the trees
themselves, and then with a soundscape composed of animals native to
the area but no longer there.

    Ghost Forest works, that is, by rising in visitors' sight lines and sonic
fields, meeting them as they are moving about their day, mixing everyday rhythms with different temporalities and knowledges. This mixing
happens sensuously, simultaneously, and it combines the directness of
elemental witnessing with the indirectness of imaginative sensation to
yield an intimate proximity. As such, the installation offers new insight
into how climate change is altering our world, demanding aesthetic responses capable of conveying urgency through means other than protest signs or if-then arguments.[6] It also draws out the importance of feeling embedded in concepts such as presence and placement.

    After a brief consideration of rhetorical scholarship on public art,
this chapter elaborates more precisely how *Ghost Forest* uses related
techniques of presence and placement to conjure magnitude, a felt quality that, as this book has shown, may well be indispensable for climate
rhetoric. To best draw out the installation's sensuous rhetoric and the
knowledges and temporalities it circulates, this chapter follows the pattern of a typical visitor's approach, first taking in the whole installation,
its title and placards, then moving in and among the trees, and finally
following the soundscape even more deeply into *Ghost Forest*.

Public Art, Public Rhetoric

Examinations of public art fall into two broad categories in rhetoric, and
both honor the field's commitment to publics and public life generally.
The first loose grouping, as I see it, consists of public memory scholars.[7] These scholars often examine monuments or memorials designed
to conduct the public business of remembering by means of material

artifacts what has been won or lost, not wholly unlike the memorial plaque to Okjökull, the lost Icelandic glacier, considered in chapter 2, but often larger and more imposing—e.g., The National September 11 Memorial.[8] The second group is composed of scholars concerned with art that is created for and presented in public spaces such as city squares, college campuses, and parks, often commissioned, designed, or placed in the name of urban regeneration. This body of work generally seeks to elucidate the democratic impulses of public art, how it enlivens urban spaces and connects people.[9]

Yet for all the rich and vibrant scholarship on art and public art from the standpoint of rhetoric, and despite, too, the growing treatment of climate change and visual rhetoric, some of which I engaged in chapter 4, rhetoric scholars have only just begun to study public climate art in particular—that is, public art that seeks to make the climate crisis more palpable, make it matter.[10] Public art can work in a more specific way than do the memorials or the urban-regeneration artworks heretofore considered by scholars in rhetoric. As the art educator Young Imm Kang Song observes, "There are many reasons why ecological public art is able to increase people's awareness and interest about environmental issues. In particular, site-specific installations invite the viewer to begin a dialogue about various local environmental issues. Public art is a gateway for community members to rethink various local issues and take ownership of an area."[11]

Sarah Douglas, the editor of *Art News Magazine*, put a somewhat finer point on the rhetorical work of public art in "Greening Public Art," a symposium hosted by Madison Square Park Conservancy via Zoom in the early summer of 2021, a few weeks after *Ghost Forest* opened. In her introductory remarks for a panel, Douglas observed that artists "coax people through aesthetic means to a place where they're receptive to thinking about [climate] issues." Doing so, she added, "involves subtlety and poetry."[12] Both Song and Douglas are talking about rhetoric, yet understandably, rhetoric's concepts and practices remain on the sidelines, as they often do when artists, critics, and educators consider how public climate art can move people. The subtlety that Douglas describes, though, is exactly what public art can bring to climate rhetoric. Public art often, as it does in the case of *Ghost Forest*, presents a kind of rhetoric that drifts and slides, a rhetoric whose line of approach might be characterized as oblique rather than direct.[13]

Lin's *Ghost Forest* combines both categories of public art with which rhetoric has so far concerned itself. First, its placement in a public urban park means that it provides what Caitlin Bruce calls a "space for encounter," though in this case the encounters that matter, as I will discuss below, happen not between people, but between people and nonhumans, within *Ghost Forest* itself. Second, *Ghost Forest* also carries memorial overtones. The rhetoric scholars Marilyn DeLaure and Lauren R. Kolodziejski each focus on what Lin calls her "last memorial," a set of interconnected permanent sculptures and an interactive website titled *What Is Missing*, all designed to help people grasp what has already been lost to climate change.[14] *Ghost Forest* intersects with this last memorial in both purpose and form, for several of the individual species featured in the installation's soundscape are also featured in *What Is Missing* under the heading "Ghost Forest—Madison Square Park."[15]

Of course it was Lin's *first* memorial, the Vietnam Veterans Memorial in Washington, DC, that initially prompted Carole Blair to urge rhetoric scholars to attend to materiality in relation to public memory.[16] Together with Lin's Civil Rights Memorial in Montgomery, Alabama, the Vietnam Veterans Memorial in Washington, DC, stands at the center of what are now classic, field-changing articles in communication rhetoric thanks to Blair and her collaborators.[17] It is perhaps fitting, then, that *Ghost Forest* provides the focus for the last chapter of a book on how rhetorical practices are responding to the unique demands of an unprecedented crisis. If, as Blair observes in her now-classic 1999 essay, memorial artworks "summon our attention to their assiduous materiality,"[18] they also summon attention *by means of* that very materiality. As the philosopher Elizabeth Grosz puts it in her meditation on art, "art enables matter to become expressive, to not just satisfy but also to intensify—to resonate and become more than itself."[19] As a work of art, *Ghost Forest*'s materiality meets and mixes with the materialities around it, intensifying sensation. As a *public* work of art, it mixes and resonates with everyday energies through a kinetic and sensuous presence.[20]

Presence through Placement

Or really, sensuous presences. To show how, specifically, *Ghost Forest* fills out what Douglas at the symposium named "aesthetic means" requires more specific terminology. Presence and placement are concepts that have circulated in rhetoric scholarship for some time now, but *Ghost Forest* highlights both their aesthetic and their rhetorical dimensions, and in doing so sharpens and even expands existing understandings of the concepts.

Rhetoric shares this commitment to presence with fine arts— indeed, as I am suggesting, installation art, because it works directly on the senses, establishes presence as a bundle of sensuous intensities. The aesthetic and the sensuous, as I have argued elsewhere, feed rhetoric in a number of ways, including the attentional, the dispositional, the imaginative.[21] Anne Frances Wysocki's perspective on a rhetorical aesthetics comes alive in the context of an artwork like Lin's: "a persuasion that follows not from a decision made inside one's mind, but from a sinew or pulse shifting, and perhaps staying shifted, in response to something meant to shift it."[22] Wysocki finds in aesthetics, that is, a way of considering sensuous, bodily responses to artfully presented material. In the deep, visceral, murky heart of feeling lies the capacity for change.

This sensuous work of presence is explicitly and frequently tagged by Perelman and Olbrechts-Tyteca with the phrase "the feeling of presence."[23] Indeed, for Perelman and Olbrechts-Tyteca, this feeling of presence is what distinguishes presence from plain existence: "It is not enough indeed that a thing should exist for a person to feel its presence."[24] Phaedra Pezzullo helpfully draws out the tangible and sensible dimensions of presence, and Michele Kennerly emphasizes the visual nature of the kinds of presence facilitated by rhetoric when she observes that "[r]hetoric's work often consists of giving presence to the *unseeable*—something not yet or never capable of being seen—or to the *unseen*—something visible but ignored."[25] Presence in art, as in rhetoric (or in art approached as rhetoric), can be conceived as the concept that draws forth and holds space, that provides a shimmer of intensity, that catches and fixes the roving senses, even if momentarily.

The lesser-studied concept of placement offers a way to consider more exactly how presence manifests in an art installation. Indeed, as Megan Poole and I have shown, Kenneth Burke honed the concept of placement with the help of a MoMA installation at the end of the Second World War. Placement, as we note there, delimits, defines, and "as a creative act . . . sets things in relation to each other."[26] Placement is especially useful in the context of public art, because the descriptor "public" is in part a matter of place—a place through which people move, and where they can happen upon an artwork. Lin's placement of forty-nine dying trees in the middle of a lush urban park classed as an arboretum, and her addition of a soundscape that places—or re-places—sounds of animals that used to roam the area, helpfully exemplifies the rhetorical work of placement. In this case, placement works as a means of sensuous presence, circulating temporalities, holding space for knowledges. To understand more precisely how, it helps to begin with the trees.

Raising the Dead

The placement of *Ghost Forest* in the geometric center of a park in the middle of the country's most densely populated city sets up a striking

5.2 Maya Lin's digital model of *Ghost Forest*, by permission of Maya Lin Studio. © Maya Lin Studio.

aesthetic relation to the surroundings. The Atlantic white cedars are tucked into the center of the seven-acre park, surrounded by curved walkways lined with memorial benches and by the park's 300 or so trees. When I first visited the installation in early May, as I walked down Madison Avenue I had trouble locating it through the density of trees at the park's perimeter. Then I took the park's side entrance at 24th Street, and just as I rounded a stately, scaly hackberry tree, the vista to the Oval Lawn opened up.[27] From that distance—probably about sixty yards—I could see *Ghost Forest* in its entirety, forty-nine Atlantic white cedars arranged randomly in a grove.

A point a bit farther around the lower part of the lawn's oval offers the perspective Lin used to create a digital sketch of the installation, shown in figure 5.2 above, and figure 5.3 is a photo of the installation that I took on the day it opened.

Viewed from this angle, the trees' wispy branches reach skyward, toward the Flatiron District's distinctive rooflines, their outlines near inversions of the Empire State Building, visible in the background. The newly leaved branch of a thriving crabapple tree curves into the frame.

5.3 Maya Lin's *Ghost Forest* from a perspective similar to the digital model. Photo by the author.

Once in full view, the Atlantic white cedars of *Ghost Forest* seem to rise up from the newly seeded emerald grass. These trees, now placed in a thriving, thrumming urban park, bear the news of their own demise. They also bear a history, an ecological significance, and, by most accounts, an overall eeriness. Viewing them in this way, surrounded by the park's other trees, themselves identified by small plaques and exuding their own species' histories, can help draw out the history of this tree species.

On each of the four sides of the installation, posts display placards that, after explaining ghost forests as "mass die-offs of once-vibrant woodlands," includes this paragraph about the installation's tree species: "In the past, Atlantic white cedars were plentiful on the East Coast, providing at least 500,000 acres of habitat for unique plants and animals before the 1700s. Today, Atlantic white cedar populations have dwindled to well below 50,000 acres and are now endangered by threats from climate change including extreme weather events that yield salt water intrusion, wind events, and fire."[28] The placard focuses on the current climate crisis, but Lin herself discusses the trees in an even broader context of settler colonialism. As she told the New Yorker, Atlantic white cedars "were first cut down for building our cities."[29] More than three centuries ago, when settlers learned of the rot resistance unique to this species of tree, they began to use them to build ships, roofs, and fences. During the Revolutionary War, they used white cedar charcoal to make gunpowder. A 1985 Forest Service report lists the species's "principal uses" as "poles, shingles, woodenware (tubs, pails, churns), and lumber. The lumber was used for planing mill products such as siding molding, and for water tanks, boat construction, boxes and crates, and fencing."[30] Heavy logging of years past combined with accelerated sea-level rise in recent years means that only about 22 percent of Atlantic white cedar tracts remain.[31] In Virginia, such conditions have landed this tree on the state's rare plant list.[32]

In his magisterial book The Pine Barrens, John McPhee pronounces Atlantic white cedars "the most beautiful trees in the Pine Barrens."[33] Left undisturbed, these trees can, by most accounts, live as long as 1,000 years, perhaps longer, and reach heights of 75 feet depending on the location.[34] Deer feed on their needled leaves. Yellow-throated warblers nest in their stands. The trees are exclusive hosts for caterpillars of one species of butterfly—the Hessel's hairstreak. According to one federal report, Atlantic white cedar trees are likely "habitat indicators" for a host of other rare plants.[35] Atlantic white cedars serve as a kind of filtration system in the swamps where they grow, absorbing pollutants and holding water for times of drought. Atlantic white cedars, in other words, are vital to their region's ecosystem.

But the very features that make these trees vital also make them vulnerable. Their thin bark and flammable leaves make them susceptible to fire.[36] McPhee describes how Atlantic white cedars remain standing, "feathery and dead," after fire sweeps through their patch of forest. As he puts it, they "commemorate the wildfire that took them."[37] Likewise, a ghost forest of Atlantic white cedars, also dead, thin branches stretching upwards, looms as a testament to the disruption of equilibrium that took it. These trees require soil that is completely saturated, but storm surges, such as the massive, unprecedented surge caused by Superstorm Sandy in 2012, can inundate the root system with salt water. In such conditions, if the water pools around the trees' roots longer than a couple of weeks, entire stands of Atlantic white cedars can die relatively rapidly, faster than their hardwood counterparts, and all at once. When this happens, they leave behind what geologists and marine scientists refer to as ghost forests.[38] Ghost forests stand as direct, visible, sensible indicators of a rapidly changing climate. As harbingers of what is to come, their starkness exerts a still, if still thrumming, potency. As the next section shows, even scientific accounts of their formation convey the haunting quality of these forests.

Ghost Science

Research ecologists working up and down the eastern US coast have documented the speed with which entire stands of trees die off. One ecologist, Greg Noe, working for the US Geological Survey, described for journalists the rapid formation of one ghost forest of cypress trees on the Savannah River between South Carolina and Georgia as something that "happened right in front of our eyes, much faster than we expected."[39] Noe goes on to call ghost forests "a dramatic expression of climate change."[40] The marine ecologist Matthew Kirwan told *Time* that "Ghost forests are the most striking indicator of climate change anywhere on the Atlantic coast."[41] This is because stands of trees die off rapidly and visibly as a direct response to sea rise. Inland ghost forests such as the one Lin and her spouse happened upon on a hike in Colorado are often caused by infestations of beetles whose populations thrive thanks to increasingly warm winters. In a scholarly article, Kirwan and his coauthor, Nathalie W. Schieder, refer to ghost forests as "direct indicators" of climate change, equating them with melting glaciers as outward signs of a too-quickly warming planet.[42] As Kirwan put it to a writer for *Time*, "You can touch it and see it. . . . It's just as real as a melting glacier."[43] Here, the observable phenomenon works "directly on the sensibility," just as Perelman and Olbrechts-Tyteca said many things do in the context of their discussion of placement. But here in the case of ghost forests, the rising awareness that results from that directness is cast in language of eeriness or even haunting.

"Forest conversion" and "forest retreat" are the blander names for the process by which a stand of trees is submerged, the trees' roots inundated by seawater, the swampy area transformed and overtaken, as seawater mixes with freshwater wetlands. But ghost forest names not so much a process as what remains. The National Oceanic and Atmospheric Administration's (NOAA) fact page on ghost forests defines them as "the watery remains of a once verdant woodland."[44] The entry explains, "As sea level rises, more and more saltwater encroaches on the land. Along the world's coasts and estuaries, invading seawater advances and overtakes the fresh water that deciduous trees rely upon for sustenance. The salty water slowly poisons living trees, leaving a haunted ghost forest of dead and dying timber. Still standing in or near brackish water, the decaying trees of a ghost forest resemble giant graying pillars that protrude into the air." The beginning of this entry is noteworthy for its departure from more typical staid scientific description. The writer bears down on the "ghost" part of the phenomenon's title by inserting a near redundancy with the word "haunted." The very name of course carries tones of abandonment—think "ghost town"—even as it conjures an eerie hush.

And so, while the scientific articles published by coastal forest experts opt for the word "striking," when the same experts speak to science journalists they select words like "eerie" or, as in the NOAA entry, haunted. Such insertions into otherwise intricate discussions of a scientific phenomenon indicate the creeping, bodily awareness that accompanies the expert witnessing of such drastic transformation. The shift into language of eeriness, haunting, and viscerality marks the early arrival of a realization, the scientific understanding of what, exactly, ghost forests portend. Such descriptive shifts might be aptly characterized as felt science, science whose implications can be eye-widening, can take on a kind of urgency, much like the learned significance of the data carried by the curves examined in chapter 4.[45] In the context of *Ghost Forest*, this category of felt science captures the sensuous intensity of ghost forests, how Lin's installation harnesses their status as direct indicators of climate change and, through placement—literally moving part of a ghost forest from one place to another, more accessible place—brings those indicators before the public senses.

Wetland ecologists and other geological scientists, through expertise and proximity, have early access to what is happening on the eastern coast of the US. Once formed, the kind of ghost forest featured in the installation is difficult to reach, let alone see. As George Zimmerman, an expert on Atlantic white cedars, put it for a *Philadelphia Inquirer* story, "Most people don't see it, and that's part of the problem."[46] Scientists often assess them from kayaks and helicopters. But physical access is not the only barrier to public awareness; scientific expertise, a data-driven understanding of what, exactly, these ghost forests portend, also presents difficulties. Data and their implications take on a felt quality that seeps in, not unlike the salt water, enacting transformation, raising alarm. As the wetland ecologist Keryn Gedan told *Time* magazine, "It is an eerie connection between sort of their visual beauty and the implications of how humans are changing the landscape."[47] That eeriness still circulates, albeit differently because the implications are perhaps less immediately accessible there, in the public art installation. Here, it is important to consider the placement of the grove in an urban park, itself classed as an arboretum.

A Forest among Trees: Aesthetics

Like most arboretums, Madison Square Park marks time seasonally. In late March, when the Atlantic white cedars were installed, among the first professional photographs released (but before the installation opened) was a dramatic photograph taken within the installation angled skyward (figure 5.4). The trees' branches edging the Metropolitan Life Building's clock tower lend the installation a gothic feel.

But that was March, before spring's full arrival. Spring's blooms and leaves introduced contrast with the dying trees. First there is a height contrast between the Atlantic white cedars and Madison Square Park's mature overstory. The park's rugged and splotchy London planes, scaly hackberry trees, and two remaining two-hundred-year-old English elms, for example, arch far above the installation. By early May, those

5.4 *Ghost Forest* with Metropolitan Life Building in the background, Maya Lin, *Ghost Forest*, 2021. Forty-nine Atlantic white cedar trees. Collection of the artist, courtesy Pace Gallery. © Maya Lin. Photograph by Andy Romer / Madison Square Park Conservancy. The exhibition was organized by Madison Square Park Conservancy, New York, and was on view from May 10 through November 14, 2021.

stately trees, fully leaved, curved and towered over the comparatively spindly and straight Atlantic white cedars, creating something of an odd insulating effect.

This contrast between the drab trees of the installation and the lush, cultivated urban landscape of the park featured in most of the descriptions of *Ghost Forest* published prior to the trees' installation. A brief November 2019 *New York Times* notice about the plans to open in June 2020 begins, "Those walking through Madison Square Park in June may find themselves wondering why some trees seem to be dying at the lushest time of the year."[48] The journalist quotes Lin: "I want to pull this stark image into the middle of Manhattan where everything is very green."[49] The Conservancy's press release points out the contrasting landscapes, even framing them as "alternatives" in this quotation of Rapaport: "'*Ghost Forest* presents two striking alternatives within the context of Madison Square Park—the ashen trees standing in contrast to the vibrancy of the park.'"[50] The critic Louis Bury used the gray-green color contrast to title his review.[51] The effect of the contrast is so striking as to even result in confusion. One woman, making her way uptown for a meeting, noticed the installation and stopped a security guard to ask if the trees were real. ("They're fake!" was his reply. They were of course not fake.) The contrast achieved through placement disrupts, and this disruption attracts attention, slows the pace of passersby, invites relations. On my visits to the installation, according to my notes, I observed this pace-slowing several times. People would slow down as they reached for their phones to snap a photo, and then look around for the nearest placard. Joggers especially would pull up to an abrupt halt, much as is described in a *New York Times* story: "Carla Murphy, a programming manager for the New York City Fire Department, was out running through Madison Square Park in October when 'Ghost Forest' caught her attention. She stopped dead in her tracks and began listening to the exhibition's accompanying soundscape."[52] One of the points of placement, then, is the mixing that happens, and how the place itself provides material for mixing in, for surrounding, for relating. And in this case, the material moves, grows, and changes.

Linger and Weave: Placement, Movement

Moving into the installation, among the trees, deepens those relations— the forest becomes the trees. As Maya Lin told the *New York Times* journalist Robin Pogrebin, "'There is something emotionally ghostly about walking through one of these forests.'"[53] And she is not wrong. A section of the brief audio tour narrated by Rapaport addresses visitors in the second person, drawing a complexity of felt presence from the trees themselves: "Walk into this work and stand in its center. Linger and weave through its perimeter. You'll experience the central and implied tension in the work. *Ghost Forest* is visually stark and emotionally meditative. It's beautiful and haunting, looming and claustrophobic. The force of those trees bears down and hovers over human scale and your physical presence. Their monumentality diminishes the viewer. The trees are the material and the message of *Ghost Forest*."[54] This passage, styled with the intimacy of the second person, invites visitors further in, suggesting styles of movement without directing that movement—to linger and weave is to drift. The movement without direction contrasts with the directed movement that Aoki, Dickinson, and Ott describe in the context of the Draper Natural History Museum. Rapaport's suggestions align more with the floating aesthetic presented by Wysocki.[55] Syntactically and aesthetically, this wandering slips visitors into the work itself, and vice versa. The suggestiveness of the imperative mood ("linger and weave") moves, in the subsequent sentences, through the predictive future tense ("you'll experience") to a descriptive assessment of the work: "*Ghost Forest* is visually stark and emotionally meditative. It's beautiful and haunting, looming and claustrophobic." The next sentence encapsulates the work's physical magnitude and presents it in terms of scale: "The force of those trees bears down and hovers over human scale and your physical presence. Their monumentality diminishes the viewer." Through their material, sensuous presence, the trees make human visitors small, creating something of a relational aesthetic that effectively right-sizes an otherwise too dominant species. Medium, message.

The sheer height of the installation's trees in this way asserts a sensuous presence that can confer magnitude, much as the outlandlishly tall ancient victory column reorganized the landscape of third century BCE Alexandria, as Jonathan Balzotti and Benjamin Crosby show.[56] But whereas the towering column exuded magnitude through its singular (imperial) dominance, the Atlantic white cedars of *Ghost Forest* tower over visitors all at once, their stark branches reaching far above the tallest of humans, casting their uniformly angled shadows in late afternoon. The dying trees loom over visitors, imposing without insisting. One of the staff members organizing the reporters at the May 10 press preview told me that Lin's original plan was to have more variation in tree height, but that plan changed when Lin realized it would increase the "presence"—this is the word the staff member used—if the trees were more uniform in height. The trees as installed range from 38 to 48 feet above ground. But the installation, as I have mentioned, is itself surrounded—"engulfed" would not be too strong a word—by the older, statelier trees in the arboretum, which in turn are towered over by the Flatiron District's multistoried buildings.

The placement of each individual tree deserves mention as well. On the day the installation opened, Lin told visitors that she spaced the trees with a density that would allow intimacy, and she prefaced that point by saying that perhaps they (she used the pronoun "we") were thinking of social distancing and COVID-19 and the need for intimacy. But the intimacy she is after is not necessarily between humans. When speaking of the trees' placement, Lin told the *New Yorker* writer Zach Helfand, "I want you to connect on a very visceral, one-on-one level with each tree."[57] Lin's placement of the individual trees followed the trees themselves. In the short documentary created and posted by the Conservancy, Lin says, "I had to kind of place each one in relation to its neighbors, and so that you'd have this dance between the trees."[58] All of the trees are around sixty to eighty years old, Lin told us, and each tree (along with electrical lines and sprinkler lines buried under the lawn), she said, helped her decide where to place the next one. One reviewer follows Lin's language of dance, noting that the trees' "careful arrangement (close enough together to form a recognizable grove but far enough apart to encourage sitting on the manicured lawn) creates a sense of choreographed grandeur."[59]

Once visitors heed the encouragement of the audio tour, or the posted signs reading "Lawn Is Open," and enter the work, moving among the trees, they may notice what I did: that the surrounding neighborhood recedes into the background, still visible and audible, but mixed. With no greenery, and with their pencil-like shape contrasting with the crooking and leaning crabapples, the trees of *Ghost Forest* do not inhibit sight, but they do stand in sight lines, often framing views of buildings.

Close up, the trees distinguish themselves even from each other. Knotted mounds dot the bark where branches once extended. Some branches remain intact, curving upward, and some are spiky and broken. The deeply ridged bark on a few of the larger trees alternates gray and deep rusty brown, displaying rich variegation (figure 5.5), and the bark on a few of the trees spirals upward from the roots, perhaps to distribute nutrition more evenly (figure 5.6).[60] Still other trees peel and splinter. Moving among them and noticing their distinctness can confirm and intensify their sensuous presence, with a powerful, meditative slowness.

These trees, visible and visibly near death, manifest the here-and-

5.5 *Ghost Forest* tree close-up. Photo by the author.

now presence of the climate crisis through their multiply sensuous presences, and they also, importantly, have the capacity to activate indirect sensation, what Aristotle called *phantasia*, the ancient approximation of the imagination. The trees, that is, carry stories of almost infinite relations—relations between settler colonists and Indigenous peoples and lands, between human and nonhuman animals, industrial relations, ecological relations, relations between land and sea, between space and time, between large storms and a region. The trees don't so much symbolize as bear those relations—hold them together for visitors to find and consider. They conjure the sensuous presence of these lives and pasts, and they do so obliquely, quietly. The trees' sensuous presence, their imposing size, and the way they engulf visitors all offer a potent mix of direct and indirect sensation. The visible, touchable, material Atlantic white cedars, raised in the middle of a park thrumming with activity and bursting with spring blooms, raise visions of past lives and worlds (the trees' and others') while also stretching into those of the future. Placement, that is, facilitates the circulation of multiple temporalities—seasonal time, crisis time.

5.6 *Ghost Forest* tree with spiraling bark and a deep "wound." Photo by the author.

Sensuous Temporality

At the press preview, Lin noted that the trees "will bear witness" to all the seasons, what she called "the life cycle of the park itself."[61] Here she explained that the trees were installed at the very end of winter, and that *Ghost Forest*'s planned duration meant all the seasons would cycle through the installation. Just before the installation opened, the park's 43 varieties of daffodils signaled the beginning of spring. Daffodils were planted in the park twenty years earlier as part of the Daffodil Project, a citywide planting effort begun in 2001 offering a "living memorial" to victims of 9/11. Then in the second part of April came the tulips. For an exhibition called "Garden Jewels," 8,000 tulips were on view in almost every bed in Madison Square Park.[62] These were followed by the blooming of the park's 35 varieties of redbuds, and then, in mid-June, the drama of the park's 350 hydrangea plants, blooming in their distinctive spherical clusters. July brought the coneflowers and crape myrtles. The trees stood through most of official autumn, cycling back to mid-November, a time that, as most New Yorkers can tell you, feels and looks like the other edge of winter. Daily temporalities, too, circulate through the installation, with the sun changing sides and shadows flipping as the toddler-time music of mid-morning eventually gives way in late afternoon to jazzy saxophone tones.

Layered in with the more cyclical temporalities of days and seasons are temporalities of crisis. Preview announcements of the work described the trees as having been destroyed by Hurricane Sandy (technically a superstorm), which made landfall near Atlantic City and wreaked havoc in Manhattan in October 2012. But by the time the installation opened, Sandy had a much more muted presence. Lin mentioned it at the press preview as the kind of climate-intensified storm surge that can destroy wetlands like those in the Pine Barrens.

Indeed, it turns out that trees killed directly by the unprecedented surge brought by the 2012 storm were, nine years later, too dead to be used safely as part of *Ghost Forest*. As I mentioned earlier, Lin needed to use trees that were not completely dead yet. So the memory of the storm, and its uneven impact, still circulated after opening, if more faintly. In contrast to the advance publicity, the installation's placards contained no mention of the storm surge that flooded the subway tunnels and hospitals in lower Manhattan and homes all over New York City, and that engorged the estuaries of the Pine Barrens. Indeed, Sandy's surge yielded in urgency and recency to another surge: the surge of the city's COVID-19 cases and deaths in the spring of 2020.

The global pandemic was still very much ongoing when *Ghost Forest* finally opened. At that point, reminders of the pandemic were everywhere in and around the installation—in its year-long delay, in the relative quiet of the city, in the masks hanging off the faces of joggers who stopped to gape; in Lin's confirmation of a reporter's vaccine status before asking her to step closer so that she (Lin) could better hear her questions. The pandemic also serves as the audio tour's implied precursor to *Ghost Forest*'s main crisis, "the collective vulnerability of the *next* looming global crisis."[63]

The *New Yorker*'s coverage of *Ghost Forest* joined the loss featured by the installation to the widespread loss wrought by the pandemic.[64] A jackhammer clanged on at One Madison to make room for a 1.4-million-square-foot office tower, in spite of the grim near-term prospects of such a project.[65] Figure 5.7 shows a warped and weathered distancing reminder, blaring red, in front of a newly installed placard describing *Ghost Forest*.

The installation's placement and its sensuous presence circulate crisis temporalities: the now muted crisis of Superstorm Sandy, the not yet fully muted crisis of the pandemic, and the creeping rise of the climate crisis demonstrating how public art's materiality meets, mixes with, and intensifies materialities already in play. In the case of Sandy and the

5.7 Pandemic time. Photo by the author.

pandemic, particular crisis temporalities circulate with a kind of ambient awareness of disproportionate impact.

Lin's installation calls attention to how presence—in this case the looming presence of dying trees—can perform the task of conjuring, of sliding the memories and imagination of indirect sensation backward and forward in time, activating visions of past and future, giving them presence, holding them all together. Once again, it is the "ghost" in *Ghost Forest* that enables the holding and circulating I have been documenting in this section. This ghost, like all ghosts, lingers, rattles, unsettles. As Avery F. Gordon observes, "The whole essence, if you can use that word, of a ghost is that it has a real presence and demands its due, your attention."[66]

For all the talk about presence in rhetoric, and all the talk about haunting, the two have yet to be brought into direct, extended relation.[67] The ample characterization of the rhetorical work of presence offered by Perelman and Olbrechts-Tyteca, discussed earlier, makes no mention of haunting. And yet a ghostly presence often manifests through feeling—chills and shivers run across or just beneath the skin. John Murphy comes perhaps the closest to a language of haunting when he uses the language of magic to discuss presence. "As a rhetorical strategy," writes Murphy, "presence possesses a kind of magical quality, one difficult to describe in discursive academic language and one that is, perhaps, best represented by the implicit metaphor in its name. An auditor 'feels' the argument; it almost seems to be in the room."[68] This magical quality approximates the felt quality of haunting, and so Murphy's articulation of presence makes way for ghosts.

Eve Tuck and C. Ree, in "Glossary on Haunting," present ghosts as concocted by ruining and by erasure, and they discuss how "haunting is both acute and general; individuals are haunted, but so are societies."[69] Viewed in this way, the ghost part of Lin's *Ghost Forest* provides a useful way to think about the multiply sensuous presences wrought by the installation. Some (like the trees) work directly on and through the senses, but most float, flash, and slide past, memories overlain through relation. Yet for all the meeting and mixing discussed so far, the installation's haunting takes hold perhaps most profoundly in its soundscape.

"Our Way In"

Ghost Forest's soundscape, an arrangement of twenty discrete recordings of animal species that used to dwell on and around the island now known as Manhattan, draws visitors further into the installation by providing a lively and haunting means of sensuous presence. The soundscape's integration of recorded wildlife sounds with the everyday soundscape of the park can yield a kind of magnitude. Sound, in the case of *Ghost Forest*, enables what sound the scholar Brandon LaBelle, in his latest book, *Acoustic Justice*, calls "bringing to presence." LaBelle uses the phrase to describe how sound can mix places, drawing together "a here *and* a there" (and, I would add, temporalities, a now and a then).[70] Sound, and works of sound art in particular, as LaBelle points out, "often bring one into contact with forms and forces of strangeness, bridging . . . oneself and another, this human figure and other species, the seen and the unseen."[71] This last pairing—the seen and the unseen—holds together the visible, haptic qualities of the dying trees, a sensuous elemental presence, the *here*, the now, with the not-here, and the then, of the animal recordings.

To access the soundscape, visitors to *Ghost Forest* were invited to scan a QR code with a smartphone (figure 5.8). The soundscape runs 12 minutes and 28 seconds, and it opens with the distinctive rising call of the Eastern whip-poor-will. From there, a one-sentence description ("This soundscape highlights the sounds of some of the native species of animals that were once common to Manhattan") is followed by a land acknowledgment, the creation of which was guided by the Lenape Center in New York City.[72]

The soundscape then moves through twenty discrete field recordings

5.8 Soundscape placard at the entrance to *Ghost Forest*. Photo by the author.

of choruses and buzzes, splashes and songs, howls and calls. The first recording presents a "dawn chorus" of sounds from a forest in West Virginia, featuring "the voices of birds that were present in Manhattan's deciduous forests more than 500 years ago."[73] After the chorus, the soundscape moves into recordings of distinctive sounds made by animals, identified first by their name in English, then the Latin species name, the Unami name, and, where available, the Munsee name. Unami and Munsee were the two languages spoken by the Lenape people who once lived in the region. The individual species sounds range from land animals (a gray fox, a cougar), to aquatic animals surrounding the island (song of the North Atlantic right whale, the slap and splash of a harbor seal), to those of the sky, the bald eagle, the barred owl, to the greater prairie chicken, offered in place of the now extinct heath hen, and to amphibious spring peepers. The final segment presents a freshwater wetland chorus. The soundscape then closes with another land acknowledgment honoring the "Lenape Delaware people, the original stewards of this land," and a pledge to continue that stewardship through specific actions.

This almost chiasmatically organized soundscape—land acknowledgment, chorus, land, sea, air, land, chorus, land acknowledgment—

was designed specifically to be listened to while moving within the installation.[74] As the park's news feature, posted on the Conservancy website, puts it, "The artist's goal for this audio and her website, What Is Missing?, is to focus public attention on and bear witness to the significant changes from massive habitat loss of species due to human intervention."[75] Here, the recorded sounds, like the trees with the seasons, bear witness through material presence.

The soundscape provides something of a sonic bestiary, that age-old (Western) genre that compiles depictions of particular and particularly affecting nonhuman animals. And although the eight images featured on the placard also evoke a bestiary, zooming in as they do on visibly identifiable features of eight select species, the soundscape itself, as this section shows, provides a profound and haunting "way in" to the matter of what is, and has long been, missing from that land—how the muting of their sounds accompanied the silencing of traditional ecological knowledges. The soundscape, that is, conjures magnitude through layered sonic presences.

The soundscape links *Ghost Forest* to the artworks Lin includes in what she has called her "final memorial," which comprises a set of sculptures and artworks in an array of locations, along with a website entitled *What Is Missing*, referred to both with and without a question mark. At the 2014 unveiling of *Sound Ring*, a sound sculpture designed by Lin as a gift to the Cornell Lab of Ornithology for its conservation efforts, the artist had this to say: "There are things that I want you to almost get to know in a more intimate personal way, and I think sound is our way in to these places and species."[76] What Lin's sound sculpture features, by means of its eight embedded speakers, is field recordings of animals, drawn from the laboratory's remarkable Macaulay Library, an extensive natural history archive of recordings, videos, and photographs of birds, fishes, amphibians, and mammals. The Macaulay Library also collaborated with Lin to produce the soundscape that accompanied *Ghost Forest*.[77]

In her remarks at the *Sound Ring* dedication, Lin invokes intimacy, as she did when discussing her approach to spacing the trees in *Ghost Forest*. *Intimare*, the Latin verb from which intimacy is derived, was a verb of motion used to name the act of putting or bringing into.[78] It is thus a relational verb, and even an immersive one. *Intimus*, the noun and adjective form, described the innermost, deepest, and most profound things. And for Lin, sound offers access to—or really, access deeper in to—the matter of species loss, which is what both *Sound Ring* and the *Ghost Forest* soundscape amplify. In her remarks at *Sound Ring*'s dedication, Lin described listening to a recording of the common loon, and how that helped her "realize how critically important it was to hear and bring these incredibly, at times, haunting sounds to life."[79]

As rhetoric scholars of sound and sound studies scholars repeatedly remind their readers, sound is itself multisensory, for it is composed of vibrations that are felt, and felt vibrations are what can result in hearing, even at times—in the case of high-intensity or high amplitude sound waves—a full-body sensation, one that can be felt on the skin, in the teeth, deep in the chest and abdominal walls. Sound's haptic mode of travel, its vibrating waves, can facilitate, that is, a burrowing into the imaginative meanderings of visitors, and even potentially into visitors' guts and minds, beckoning them deeper into the installation itself. The immersion is multiple, its magnitude profound. For rhetoric, the soundscape shows how rhetoric is transduced as energy rather than transmitted as meaning, as scholars such as Casey Boyle show in the context of new media.[80]

Going deeper into the material composition of the soundscape—the field recordings that are assembled there—can show how, in particular, this multiply sensuous presencing works. As Michael Gallagher, a music producer and social researcher, explains in an article on field recordings, "What the field recordist brings home is not sound from the environment, but arrangements of charged particles in the semiconductive materials of solid state 'flash' memory, or the magnetic surfaces of hard drives, tapes, and minidiscs. These recordings are not sounds. The sounds, if they are anywhere at all, are still 'out there' in the field, dissipating through space."[81] The recording is made into sound anew when this electrical energy is compressed and converted to an .mp3 file, which is then converted back to sound waves and into what Gallagher calls "space of playback."[82] What Gallagher makes plain here is the particular quality of sound, which is to say that what we call sound is modulating, energetic particles that move in waves. Those bundles of energy, disturbances in the air, are out there, scattering, a snippet of their vibrational patterns now transduced into data, compressed and arranged into an .mp3 file, and in the case of *Ghost Forest*, accessed through visitors' phones or other electronic devices, reconverted to air, directed to ear.

A Park Avenue hotel room was my first space of playback, the device a MacBook Pro. The first tones that issued forth from my laptop brought with them a flash of recognition from the woods of eastern Tennessee, where the whip-poor-will provided the backdrop of my childhood summers.[83] Right behind the whip-poor-will in this playback space came the electronic carillon bells from the nearby Church of

Our Saviour chiming nine a.m., moving through an open window, their deep tones sliding over the comparatively tinny soundscape. Gallagher's overview of the soundscape's material and energetic forms parses the material and temporal mixing that can result when these bundles of acoustic energy, these disturbances, are played back by visitors while moving through the installation. On my first listen, the soundscape's compressed soundwaves conjured and mixed with humid July evenings, shimmering fireflies and trilling frogs—the then—and with the digital tones of church chimes marking the morning hour—the now. This 45-second stretch of attentive listening brought home for me sound's temporal latitude, its ability to hold the flashes with the streams, to slide worlds and times across each other, opening imaginative pathways: ways in.

What Lin knows as an artist about the medium of sound as "our way in," the media theorist Francis Dyson asserts emphatically: "sound," writes Dyson, "is the immersive medium par excellence."[84] As an immersive medium, sound reduces distance, circulates and recombines temporalities. Sound, that is, holds space and composes presence. In the case of *Ghost Forest* in particular the soundscape creates an immersive magnitude through mixing and through haunting, both sensuous techniques of presence.

Later that morning, Lin's *Ghost Forest* became for me the second space of playback, the space the soundscape was designed to enliven. This time, my iPhone converted waves from the .mp3 file to ultra-high-frequency radio waves, transmitted via Bluetooth signal to my Airpods, and as I moved through the installation, the calls and voices vibrated inside my ears, meeting and mixing with the sights and sounds of Madison Square Park. The jackhammer noise coming from the southeast led me to increase my phone's volume and move toward the installation's northwest corner. On the way, the soundscape's recorded howls of gray wolves mingled with the yips and barks of the park's dog run. Another visitor turned up the soundscape volume on her phone so that the distinctive opening sounds of the whip-poor-will mingled with those emanating from the lawn to the north of the installation, where a face-shielded ukelele player entertained a group of toddlers with the energetic song "We are the dinosaurs marching, marching": "And then we ROAR!" One member of the singer's audience emitted a chesty wail from a stroller.

A similar sort of mixing features in Marilyn DeLaure's analysis of yet another of Lin's sound sculptures, the *Listening Cone*, a permanent piece installed at the California Academy of Sciences. DeLaure writes,

"My experience of the Listening Cone was not a 'pure' sonic one: for much of my time sitting inside the sculpture, other museum guests and staff were chatting in the picnic area behind the cone, and the low hum of city traffic remained constant."[85] For DeLaure, "The Listening Cone created a zone of friction between the immersive natural soundscapes inside the Cone and the everyday sounds of the human world, mirroring the incursion of anthrophony into the biophony (natural sound) that is one concern of soundscape ecologists."[86] DeLaure's point about the friction generated by the layers of sound is especially apt in a museum context, and especially, too, in the case of *Listening Cone*, which is itself designed *as* an acoustic device, a space of playback. But the *Ghost Forest* soundscape differs in that it was designed to guide visitors through an urban space, a space that does not feature a natural history museum's norms of directed movement.[87] When *Ghost Forest's* soundscape is played back in the middle of Madison Square Park's cacophony, the sounds that dominated on this land centuries ago slide across and mix with the sounds of wailing toddlers, barking dogs, and sirens. *Ghost Forest's* soundscape, that is, makes the space anew, "generating vibrating fields of waves" among vibrating fields of waves.[88] The sensuous, sonorous presence taps imaginative capacities, introducing a profound temporal mixing.

The mixing of sound brings me back to haunting, which I discussed above as a felt presence. The soundscape, in raising the voices of humans and nonhumans now gone from a place, layers in with the installation's title and its muted grays and browns, set off, as I discussed earlier, by the lushness of the park's canopy, to conjure multiple presences. The temporal layering of *Ghost Forest's* soundscape helps demonstrate more precisely how haunting works as a technique of sensuous presence in the context of the climate crisis. In particular, the soundscape, in giving presence to people and nonhuman animals who have been displaced, even destroyed, from "the land beneath your feet" (to use Lin's arresting phrase) brings forward the horrors of settler colonialism and also connects those horrors to those of the climate crisis.[89]

Indeed, haunting, as Eve Tuck and C. Ree show, gives presence to the underside of settler memory (known, by definition, as history), to the brutal horror of settler colonialism. The soundscape is gentle with this presence; the presence lingers, though, in the Lenape voice reading Lenape names of species gone from the area, now more widely known by the names settlers called them, and particularly in the missing Munsee names, themselves dwindling along with the Munsee people and culture.[90]

Munsee names of the right whale, the gray wolf, the cougar, and so on have vanished. They are gone like the sounds of the heath hen, a bird that was once abundant in the area. Settlers overhunted the bird, and by the late 1920s efforts to revive the species were compromised first by fire and then by disease. Flickering video footage taken on Martha's Vineyard around 1927 shows some of the last known heath hens pecking and strutting in an open field.[91] *Ghost Forest*'s soundscape marks the loss of the heath hen by including a recording of a close relative whose numbers are also decreasing, the greater prairie chicken, as "stand in for the extinct Heath Hen."[92]

The Lenape words, and the living Lenape people who sustain the language, bring to presence the displaced Lenape people and culture. For example, the name *òpalanie*, what Lenape Indians called the bird that settlers call a bald eagle, carries a relation of respect embedded in traditional ecological knowledge.[93] Or really relations: between the Lenape people and the settler colonists who forcibly took their land and who also brought the heath hen to the brink of extinction. Those relations haunt the soundscape by the sounds of the animals, and again by the reading of their Unami names. The soundscape's haunting works as a necessary, persistent mode of remembering, or what Tuck and Ree call "the cost of subjugation."[94] And the rich but spare vocal tones presented by these recordings—the resonant howls of the gray wolves especially—conjure images of the creatures themselves, some too distant to be more than an imagined outline, roaming the island, swimming its waters, hunting its forests. The soundscape concocts haunting by conjuring sounds of what is missing, and also by subtly acknowledging those for which sounds themselves are missing, no longer available, their sounds still imperceptibly dissipating in that very place. The haunting becomes an act of witnessing, voicing a long-standing and still unmet demand for a reckoning.

Avery F. Gordon presents haunting as "a constituent element of modern social life," and I would argue that it is also a constituent element of climate rhetoric.[95] As Gordon shows, and as Tuck and Ree elaborate, haunting encourages a deep attunement. In a foreword to the second edition of Gordon's influential book *Ghostly Matters*, Janice Radway writes that Gordon presents "a knowing that is more a listening than a seeing, a practice of being attuned to the echoes and murmurs of that which has been lost but which is still present among us in the form of imitations, hints, suggestions, and portents."[96] Radway's phrase "among us" stands out when considering the soundscape's rhetorical form of haunting, how sound waves move and mix, slide and layer with and among other sound waves. Haunting, that is, encourages a slow and deep attunement to the climate crisis, holding together past losses with creeping portents of future ones.[97]

Conclusion: The Presences of Public Art

Ghost Forest shows the potential of public art—art in public, everyday spaces—for climate rhetoric. Specifically, this chapter, moving by turns deeper into the installation, documents how placement and haunting can work as techniques of sensuous presence for public art. This installation does not present a direct message. Rather, the installation's urban park setting places before the public the eerie, sensuous force of felt evidence. Ghost forests of Atlantic white cedars, heretofore mainly visible and sensible to experts studying the forests as ecological phenomena, placed in the middle of a public park, reduce the distance of this direct indicator from those who live in and visit the city. The artful ("choreographed") spacing of the trees themselves and the transformation of the public park into a space of playback for the soundscape combine to show how placement and haunting work in coordination as techniques of sensuous presence for public art. When Lin's installation closed as planned on November 14, many of the Atlantic cedars were donated to a nonprofit organization in the South Bronx called "Rock the Boat," where teenaged apprentices learn the art of boat-making. Lin was delighted with this afterlife for her work, calling the boats "part of a new life for the artwork."[98] There is a poignancy to this new life: as boats, these Atlantic cedars of *Ghost Forest* might well float on the very water that hastened their demise. The haunting does not stop.

Ghost Forest brings to the unprecedented climate crisis a powerful aesthetic emphasis, bearing "emphatic witness," to borrow an arresting phrase from Wai Chee Dimock,[99] to a mix of evidence and knowledges. As an artwork, it works primarily through felt registers—the senses, feeling. The installation therefore helps expand the particular kind of magnitude conveyed by curves in chapter 4, even as it augments the youth rhetors' emphasis on felt time drawn out in chapter 3. And it does so by honoring and acknowledging traditional ecological knowledges, listening to them while also marking their silencing through haunting, an important technique of sensuous presence for rhetoric. The installation, that is, without a didactic, information-only message, nevertheless tilts toward the matter of justice. It does so by marking absences of and through Indigenous voices, knowledges, and names and by layering somber tones of loss in with the carrying on of urban public life. At times, the enormity of the task of addressing the climate crisis becomes overwhelming, incapacitating. Providing yet more information can be as insufficient as increasing the volume of or the number of voices shouting. And while it may still not be enough, the power of public art to hold and mix together multiple sensuous presences, knowledges, and temporalities—to bring, that is, that bundle of felt evidence before the public—is crucial for deepening public knowledge of the climate crisis as here, now.

CHAPTER 6

Epilogue

FATHOMING

For my forty-fourth birthday, my spouse gave me a copy of an oblong photo essay book titled *The Oldest Living Things in the World*, by the artist Rachel Sussman.[1] Although it wasn't intended this way, the gift struck me as at least partly a consolatory joke: I'm not *that* old. The book is in fact a beautifully designed tribute to geologic time. It is also a subtle meditation on the limits that lifespan—particularly human lifespan—can place on the imagination of time and especially of futures. Read in the context of this book, *The Oldest Living Things in the World* becomes an archival, aesthetic effort to pry the concept of longevity away from human lives and present it with a renewed expansiveness. It stretches the art and act of witnessing to encompass past, present, and future, humans as well as nonhumans.[2]

Sussman's book, in other words, is of a kind with the interventions I have considered in this one, efforts to stretch and strain imaginations by presenting, by making present, different magnitudes—of loss, of feeling, of mathematics, of, again, loss. Sussman's book brings forth a kind of magnitude of continuous life, of sheer duration, years piled on years, time deeper than most are accustomed to thinking. Sussman writes in the introduction, "as with deep water, it's a battle to stay in deep time," and yet her hope is that the book will help its readers tread there a little longer.[3] The collection of photographs and essays helps its users practice sliding back and forth in time, holding open multiple timeframes: "Perhaps," writes Sussman, "looking through the eyes of these ancient beings, and connecting with the deepest of deep time, we can borrow their bigger picture and adopt a longer view."[4] The attribution of eyes to, for example, clonal moss might best be read not as an anthropomorphic urge, but as a poetic form of witnessing, one rooted in enduring, deep presence.

Sussman holds the "shallow time" of her medium, photography, in tense relation with deep time.[5] "What does it mean," asks Sussman on her artist site, "to capture a multi-millennial lifespan in 1/60th of a second? Or for that matter, to be an organism in my 30s bearing witness to organisms that precede human history and will hopefully survive us well into future generations?"[6] Sussman thus presents her photography as an act and art of witnessing, one that provides access to the magnitude and profound presence of continuous life. But she is not the only witness. Indeed, in the book, she figures her photographic subjects as witnesses:

> The oldest living things have borne witness to the entirety of human history. The wheel and cuneiform in Mesopotamia—inventions that mark the birth of civilization—emerged about 5,500 years ago, the same age as the Antarctic moss on Elephant Island. Anything previous is prehistory. The 13,000-year-old Palmer's oak in the industrial sprawl of Riverside, California, lived through the die-off of giant reptiles, birds, and mammals that seem more like the stuff of science fiction, such as giant condors, mastodons, saber-toothed tigers, and even the last of the camels that once roamed North America.[7]

With a descriptive alignment of timelines, Sussman advances the kind of elemental witnessing that this book has been following, especially in chapters 2 and 5.

Many of the "ancient beings" showcased in Sussman's book are trees.[8] There is the famed Fortingall Yew located in Perthshire, Scotland, estimated to have lived for more than 2,000 years, and possibly as long as 5,000; and the Antarctic beech tree in the dense and mountainous rainforest of Queensland, Australia's Lamington National Park.[9] Pictured in the Giant Sequoia section are two of the four "King Canyon elders." All four trees, and likely scores of others, pre-date the year 0 CE, some by more than a century.[10] The baobabs that Sussman photographed in South Africa grow with age to an astonishing bulk, allowing them to store water in their expansive hollow for times of drought. Some baobabs expand to a diameter of fifty meters, which is about half a soccer field. Such measurements are difficult to fathom.

"Fathom" is itself a curious word, a word of limits, particularly human ones. It usefully holds the kinds of intensifications that I have documented in this book, conveying as it does a kind of straining. This book has examined striking instances of how people are helping other people fathom the climate crisis. Doing so means making evidence of a rapidly changing climate less abstract. As I put it earlier in this book, as an intensifier, rhetoric moves beyond presenting findings to *finding presence*, drawing time and place close enough to take them in, and entwining facts with feelings. It means making more directly sensible what Indigenous populations and scientists have long known; it means combining different forms of witnessing to make more accessible the *magnitude* of the situation, as illustrated in the memorial service for a glacier (chapter 2), testimonies of young people (chapter 3), the adaptation of the epi curve's graphic form to the climate crisis (chapter 4), or the placement of dying trees in the middle of an urban park (chapter 5). Laying bare these changes and adjustments, as I have shown, has meant using concepts that themselves bear ethical weight, or whose ethical weight becomes heavier and more discernible in the instances the book examines. Those changes may, in short, help people to fathom the magnitude of the crisis before us.

As a verb, "fathom" usually suggests attaining understanding, though as its history reveals, that understanding is enabled by reach, "to encircle with extended arms," as the *OED*'s first entry puts it.[11] The word "fathom" came to be used as a noun, indicating a measurement of around six feet, the length of a man's outstretched arms, as in a fathom of firewood, or a line measured and counted out in hand-over-hand units that can be turned vertical and dropped into the sea. The word's nautical history as a unit of measurement—British Admiralty charts replaced the term with meters in 1968—yielded its figurative associations with depths, with peering into them.[12] That which cannot be fathomed names that which cannot be embraced, or something too deep, too dark to quite make out.

The physical, sensory assessment behind fathoming is as important for its work in rhetoric as its dependency on, and presumption of, a human bodily shape. It is one among many literal instantiations of the famous dictum uttered by the sophist Protagoras that "man is the measure of all things," a dictum that might now be read retrospectively as bearing the weight of the Anthropocene.[13]

Fathoming has a history with trees too. Materially, its span from fingertip to fingertip was used in both noun and verb form to describe and measure tree girth. Writing in 1828, for example, Sir Walter Scott recounts in his journal a breakfast conversation with the landlord and agriculturalist Thomas William Coke, who "talkd of trees which he had planted as being so thick that a man could not fathom them."[14] The word appears centuries earlier in a well-known English translation of Spanish settler colonial writings as part of an account of people who built their houses in "trees of maruelous height": "Owr men measuring manye of these trees, founde theym to bee of suche biggnes, that seuen men, ye sometimes eight, holding hande in hande with theyr armes streached furthe, were scarsely able too fathame them aboute."[15] In the act of fathoming, words, graphics, art—the material stuff of rhetoric— work like arms, reaching, stretching, sometimes edging us closer to contact. Also like arms, they often come up short.

The poet C. D. Wright, in a book-length appreciation of beech trees, pauses to reflect on the near unfathomability of another specific tree: "Then there is the Montezuma cypress, El Árbol del Tule, in Santa María del Tule, Oaxaca. Words don't come close. Never have. Never will. Were I a diviner of children's books, I would compose a multigenerational Gongorismo chronicle from the point of view of the creatures who have burrowed, nested, and sported in the shelter of that spectacular tree, somewhere between 1,400 and 3,000 years old (opinions differ). I would have to consume many a magic mushroom to touch the wonder."[16] Wright's rhapsody over this particular tree captures the difficulty, if not impossibility, of fathoming El Árbol del Tule—its temporal and physical expanse, and the astounding ecologies those expanses have long nurtured. Wright also renders the urge to reach for what she calls wonder, despite the impossibility, and to require assistance in that reaching. Interestingly, Wright, writing in the subjunctive mood— considered in chapter 2 of this book as vital for imaginative reaching— finds that assistance in audience (child readers), the elaborate, poetic style of *gongorismo*, and hallucinogens—all means of stretching beyond the ordinary.[17]

Wright's observation that "words don't come close" conveys, somewhat paradoxically, the complex and intense feeling that trees carry and prompt, be they individual trees, trees in a grove, or even a species or entire forests. In that same spirit, I offer a brief meditation on trees to widen chapter 5's focus on one grove of a particular species of tree to the entire tree family. Trees provide a way to revisit the points made throughout this book: they both carry information and attract feeling; they bind generations; they fashion visible data to help us better comprehend just how unprecedented our climate emergency is; and finally, they can exert a magical, even sacred presence.

Trees, then, in all their living material wonder, assist with the kind of imaginative stretching, including the stretching beyond the human, that climate rhetoric needs. In their status as elemental witnesses and also as sacred objects, particularly in the context of Indigenous lives and knowledges, trees can help us begin to fathom different futures, a capacity that has too often been limited by human lifespans. In this way they help me to reiterate this book's points about time and temporality and the need to pry both open, to expand them to include greater varieties of witnessing and presence, and to come to terms with the climate crisis's magnitude. For some scientists and climate activists, trees even provide an important part of the solution to the carbon problem. And so to conclude, I propose to dwell—through perhaps reach is a better verb—with trees.[18]

Tree Time

In approaching trees from the point of view of rhetoric, I join a handful of scholars who have documented relations between trees and rhetoric.[19] Most rhetoric scholars writing about trees seem keenly aware of how tree timescales have the potential to expand imaginations beyond individual human lifespans. As Jason Edward Black puts it, for example, "ostensibly, trees help humans grapple with the topography of our lifespans."[20] And Barnett finds in the ecologists' writings the invitation "to think like a forest," which is to say, "to think on timescales that vastly exceed that of human life."[21] Unlike humans, as the dendrochronologist Valerie Trouet puts it, "trees can live for millennia."[22]

To fathom a tree's girth, then, is to come into contact with its life, its longevity, as Sussman and Wright do with photographs and words, respectively. After a certain age, most tree species cease to grow taller, but they continue to expand outward year by year, developing new vascular tissue patterns (or in the case of conifers, cellular growth).[23]

"Tree" and "time" of course form the two parts of the word "dendrochronology," the branch of science that scrutinizes and compares annual growth rings on trees for the purpose of measuring time's lapse. But that science tells much more than time, as Trouet's popular book *Tree Story: The History of the World Written in Rings* suggests. Working with what Trouet calls "ring by ring precision," dendrochronologists can coax information—stories, Trouet likes to call them, true ones—of past climatic patterns and movements from the rings' composition.[24] The same wildfire smoke that creates the dazzling, disconcerting atmospheric optics shown in the book's frontispiece will years from now be detectable in tree rings throughout North America. Much like the layers of glacial ice considered in chapter 2, trees serve as elemental witnesses to climatic patterns and movements. Indeed, as Trouet reminds readers, tree ring data and ice core data together provided the proxy data on which the climate scientist Michael Mann based the famed hockey-stick graph, discussed in chapter 4.[25]

It is also worth acknowledging that trees are often presented as at least partial solutions to the runaway carbon problem. Greta Thunberg and the journalist George Monbiot promote tree planting in a widely circulated video.[26] After Thunberg met with Canadian Prime Minister Justin Trudeau during her trip to North America in 2019, Trudeau pledged to plant two billion trees over the next ten years.[27] Trees are crucial parts of programs designed to gird urban areas for rising global temperature, to address yet another form of built environmental injustice of the kind that Chris Suggs, Jamie Margolin, and Aji Piper elaborated in their testimony before the US Congress. In short, the presence—or absence—of trees can entail social inequality.

Trees and reforestation are, to be sure, often the form that carbon offsets take.[28] Maya Lin's *Ghost Forest*, as one example, includes a carbon offset plan to plant 1,000 trees in New York City's five boroughs.[29] Researchers continue to refine predictions of how much carbon trees can take in.[30] Suffice it to say that the future promises lots of tree talk among humans. But will it involve listening to trees themselves? And to those who hold traditional knowledges based on such listening?

Forest Listening

The forest research of Suzanne Simard, who documented the chemical exchange of signals, nutrients, and information among trees through "a world of infinite biological pathways," has prompted a renewed popular interest in trees.[31] Richard Powers's 2018 Pulitzer Prize–winning novel *The Overstory*, which features trees as protagonists, includes a forest-scientist character based on Simard.[32]

It is important to note, though, that the deep and intricate communication networks Simard documents are not likely to surprise people who have long known how to listen to trees, whose relation to trees has for generations been balanced and reciprocal. A collaborative account of forest listening offered by Jeff Grignon, a Menominee Tribal Enterprises regeneration forester, and the environmental biologist Robin Wall Kimmerer conveys the ethic of balance that inheres in so many traditional ecological knowledges still in circulation but insufficiently heeded. As Grignon and Kimmerer describe it, the Menominee Forest in what is known now as northeastern Wisconsin holds deep knowledge of Elder Tree Nations, Elder Maples, Pines, knowledge that protects and sustains the forest and, in turn, the people.[33]

Veneration

A quality of relation can be discerned in the act of veneration, as shown in chapter 2's account of the epideictic rhetoric that emerges from efforts to memorialize the lost Icelandic glacier. Long the subject of veneration, trees, particularly old trees, large ones, or forests deemed significant, hold for many a quality of relation. "Venerable trees" lie at the heart of a nineteenth-century book made by the English engraver and painter Jacob George Strutt. The book, titled *Sylvia Britannica; or, Portraits of Forest Trees, Distinguished for Their Antiquity, Magnitude, or Beauty*, was first published in 1822, followed by an expanded version in 1830. Strutt opens the book with these words: "Among all the varied productions with which Nature has adorned the surface of the earth, none awakens our sympathies, or interests our imagination, so powerfully as those venerable trees which seem to have stood the lapse of ages—silent witnesses of the successive generations of man...."[34] In his era's signature overwrought prose, Strutt enumerates the ways that trees can capture attention and engage imagination. The etchings in Strutt's book—witnesses of witnesses—underscore his means of selection as indicated by the book's subtitle: age, magnitude, and beauty. Indeed, the fact of duration (age) combines with aesthetic pleasure (beauty) to yield magnitude, the middle term, which conveys intensity of feeling, a move I have been charting throughout this book. Strutt's adjective "venerable" reveals his book's main purpose, which is to praise trees. To venerate individual tree species, Strutt provides sketches and literary excerpts, and generally expounds—and I do mean expounds; the section on the oak runs fifty-seven pages—on the virtues of the oldest trees around him. One of the virtues that Strutt draws forth has to do with planting. A person who plants an oak knows they won't live to see the grandeur of an expansive old tree, but they know someone will. For this sentiment, Strutt amplifies Cicero: "He who puts a sapling into the ground, is morally certain that he shall not live to enjoy it every day, and a thousand fold, in the thought, that the land, which to his predecessors had been only a barren waste, will present to his successors a scene of waving beauty, sheltering the surrounding country, and inviting many a devious step to explore its tangled haunts."[35] The young tree, that is, holds a future to which the planter and tender may enter "by proxy." Much as trees provide proxy data to the past, then, they can also help conjure futures, aiding imagination through a conception of presence. And in doing so, as I have been arguing in this book, they provide forms of elemental witnessing.

Sacred Trees

The tree perspectives discussed so far—as proxies for pasts and futures, forest listening, veneration—all gesture toward a long history of trees as sacred beings. The sacred, as the rhetoric scholars Michael Bernard-Donals and Kyle Jensen present it, simultaneously provides a limit and an opening.[36] Both are what is needed, and both are what this book has foregrounded.

First, the opening—to futures unknown, to that which exceeds humans, as trees do, by many measures. The sacred, too, often entails a kind of animacy. Robin Kimmerer, in reflecting on the difficulty of the Potawatomi language, a lively yet threatened language that had been assimilated out of her life before she ever learned it, explains the animacy that courses through the language itself, carrying with it ethical relations:

> The animacy of the world is something we already know, but the language of animacy teeters on extinction—not just for Native peoples, but for everyone. Our toddlers speak of plants and animals as if they were people, extending to them self and intention and compassion—until we teach them not to. We quickly retrain them and make them forget. When we tell them that the tree is not a *who*, but an *it*, we make that maple an object, we put a barrier between us, absolving ourselves of moral responsibility and opening the door to exploitation. Saying *it* makes a living land into "natural resources." If a maple is an *it*, we can take up the chain saw. If a maple is a *her*, we think twice.[37]

The sacred, then, entails regard, veneration, the lively relation of kinship, a charge. The sacred makes present just the kind of concentric temporalities that this book has documented in moments when climate change has altered rhetoric.

In other words, the sacred, in all its liveliness, can hold open, but it also holds a line, or a limit.[38] The limit itself can open to flourishing, as in the case of sacred groves, for example. The thing about the sacred is that it mitigates against one-directional disturbance—think clear-cutting in the case of forests. Because their sacred status prohibits human disturbance, sacred groves have become sites of conservation and evident biodiversity.[39] Sacred groves and sacred trees, in other words, reveal worlds past and desecrated even as they show paths for regeneration. Sacred trees therefore join the examples of the climate rhetorics considered in this book in that they open to a broader conception of rhetoric as an art of intensification, one that holds together quality, time, magnitude, and knowledges and gives them presence by enlarging them, drawing them close.

Closer

Extending from Lin's *Ghost Forest*, this epilogue has lingered on trees—how they hold information, knowledges, and time pressed into dense rings and durable lives. Because trees can stretch human imagination and sensation beyond the ordinary, because they witness, humble, and portend directly and obliquely, and yes, because they subsist on carbon, they provide a sturdy way to gather the theories of rhetoric that this book documents. Trees, like humans, have not known a planet like the one we currently inhabit. Trees, like humans, are changing in response to rapidly changing conditions.

It should be apparent by now that this book will not end with a bulleted list of ways to address the climate crisis. For one thing, those already exist, and the IPCC updates them regularly. As the memorial plaque for the Icelandic glacier asserts, "we know what is happening and what needs to be done." Nor do I pretend to fully resolve the care crisis that exacerbates the climate crisis, though the matter of care tracks more closely with the disciplinary aims of rhetoric as an art. Instead, this book set out to see how rhetoric is changing in response to urgency escalating all around. I wrote this book, that is, to take the pulse of rhetoric at a time when the past—the usual way of looking toward the future—so often seems of limited use, and at a time when so many presumptions about persuasion, about information, about awareness are being complicated, if not overturned. Awareness and information have proven to be necessary but still insufficient for action, and the novel climate rhetorics considered in this book respond to that insufficiency by moving and changing, humbly, yet resolutely. The ceremonial rhetoric held by Howe and Boyer, the youth rhetors' presentation of felt time, the adaptation from a faster-moving crisis of a phrase and an image, and the haunting installation of dying trees together show that—and how—witnessing, presence, and magnitude are more vital than ever to the complicated work of imagining futures together.

Rhetoric, as a means of intensification, binds together fact and feeling, time and sensation, as the departures from more ordinary rhetorics show. New forms of witnessing are called forth and joined together with existing ones. As an intensifier, rhetoric bends and stretches, holds open and disrupts. Rhetoric's means of intensification involve a quality of attention more than the fact of it. These means of intensification wrap facts with feeling; they hold together temporalities near and far; they make evidence present in such a way that it can be seen and also deeply and keenly felt. And, as with trees, fathomed.

ACKNOWLEDGMENTS

Acknowledgment One

I wrote most of this book on the original homeland of Susquehannock/ Conestoga people, near the main campus of the too gently named land grant university of Penn State. The surveying, measurement, extraction, and exploitation of settler colonialism lie at the base of the confounding social, geological, and ecological problems that have long needed to be addressed. And so while acknowledgment now is not, and will not be, sufficient, it is nevertheless important to heed and honor the longer history of these lands and their traditional caretakers. That caretaking— and its violent interruption—constitutes part of the theory of temporality that this book unspools.

Acknowledgment Two

This book found its beginnings in the question-and-answer session of a 2016 keynote lecture I delivered in Uppsala, Sweden, for the International Rhetoric Workshop on "Deliberative Imagination and Democratic Life." My aim in that lecture was to inject the overly rational theories of public deliberation with more robust rhetorical theories of imagination and feeling, and I defined deliberation as a process of imagining futures together. Dilip Gaonkar, another plenary lecturer at the event, wondered from the audience if the connections I was making would find greater potency if they were grounded in a specific issue—he asked me in particular about climate change. Our speculative exchange on that question generated energy in the room, in me. His was the kind of question that lingers and grows, and it stuck with me as I continued to read, teach, and write about imagination for the next three years before finally returning to the idea full force. The result of Gaonkar's question is this book.

Vanessa Beasley and Cara Finnegan, exemplary leaders, kickass scholars, and dear friends, first heard about this book when it was a slip of an idea, and they asked encouraging and clarifying questions from the beginning. Elizabeth Mazzolini and Amy Allen traded writing updates with me and offered friendship and humor. Elizabeth shared with me her expertise in environmental humanities. My friendship with Michelle Frisco took root during this book's development, and she checked in with me at every turn. Neighborhood walks and talks with Michele Kennerly, especially, but not only, during the pandemic, provided sanctuary.

Kennerly also helped me improve the introduction. My Penn State colleague Brad Vivian and I found ourselves working with rhetoric involving children and youth around the same time, so naturally we talked about those challenges and traded drafts. His work and his feedback helped make this book better, especially chapter 3 on youth rhetoric and felt time, and he, along with an anonymous reviewer, encouraged me to go further with witnessing. Casey Boyle read chapter 5, the epilogue, and the introduction with an energy I hope to someday return in kind. Christa Olson and the members of her 2020 graduate seminar, "Recent Rhetorical Theory and Today's 'Wicked Problems,'" along with the rhetoric scholars G. Mitchell Reyes and James Wynn, helped me clarify what I am up to in chapter 4. Joshua Trey Barnett arrived at Penn State while I was in the throes of drafting, and though the pandemic prevented us from really meeting until I was nearly finished, I gratefully acknowledge his work in climate rhetoric and his collegiality in general. (More meetings that are walks in the woods, I say!) Other colleagues, too—Kirt Wilson, Stuart Selber, Mary Stuckey, Cheryl Glenn, Keith Gilyard, Jack Selzer, Pamela VanHaitsma, Xiaoye You, Steve Browne, Jeremy Engels, and Rosa Eberly—inspired me in their own brilliant ways. Claire Bourne and Claire Colebrook helped me build strength.

Graduate students who enrolled in my seminars on Imagination and on Crisis Rhetoric at Penn State, and especially my Crisis Rhetoric work sessions group, Matt Parnell and Sierra Parker, served as sounding boards for material in chapters 4 and 5, as well as the work on time that winds through the book. Regular conversations about writing and research in rhetoric with graduate advisee-colleagues infuse these pages. They include: V. Jo Hsu, Sarah Adams, David Maxson, Megan Poole, Curry Kennedy, Michael Young, Lauren Beard, Ismael Quiñones, Philip Grayson, Genevieve Gordon, and Sierra Parker.

The Write on Site scholars were virtually present for a large portion of this book's writing. Thanks especially to Stephanie Kerschbaum and Jenn Fishman for keeping the site going, and to Jessica Enoch, Katie

Gindlesparger, Annette Vee, Kate Vieira, and Amy Wan for showing up to write, even—especially—during those long, dark mornings of 2020.

Penn State University's College of the Liberal Arts and English Department, and an endowment created by Ted and Tracy McCourtney, supported this book by funding release time and research assistants. I am grateful to the McCourtneys for their belief in the research and teaching that faculty members do.

Special thanks go to Brooke Kamin Rapaport and her colleagues at the Madison Square Park Conservancy for generously inviting me into the *Ghost Forest* press session on opening day and for their help with images and permissions. Thanks too to the youth rhetors featured in chapter 3 for using their voices to speak time. Let's all listen more carefully to them, please.

Abigail Fourspring provided invaluable research assistance for chapter 5 and helped me organize images in chapter 4. Sierra Parker provided crucial endgame permission and image assistance. Steven Schneible gathered information early on about youth climate rhetors.

Kyle Wagner at the University of Chicago Press believed in this book from the beginning, and the professionalism and expertise of Wagner and his colleagues impressed me at every turn. The anonymous reviewers helped me improve the book by showing me what I was—and, just as importantly, was not—doing and helping me find ways to do it better.

John Marsh, ever my first reader, helped me figure out how to sharpen things for all subsequent readers. Kim Marsh, *ma belle-sœur*, epidemiologist extraordinaire, taught me about the sheer beauty of information more than the elegant books of Edward Tufte could. Other family members—my parents, my sister, and my family-in-law—sustain me from afar. This book is dedicated to my daughter Nora, niece Sarah Grace Peterson, and nephew Seth Bowers. Each one has connected to the book's purpose through their curiosity, lived experience, and expertise. My love and respect for them strengthens my opposition to hope-in-future-generations talk that too often stands in for action.

NOTES

Chapter One

1. This particular tree has become known as the "Dome Comb tree" thanks in part to Gravett and his collaborators, Adam Ozment and Adam Williamson. Ecologists have set the mortality rate of Fraser firs on some Appalachian mountains as high as 90 percent. See, in particular, Allen and Kupfer, "Application of Spherical Statistics," 482–93; Allen and Kupfer, "Spectral Response and Spatial Pattern," 59–74.

2. Intergovernmental Panel on Climate Change, "Climate Change Widespread, Rapid, and Intensifying."

3. The phrase "global weirding" appears to have originated with Lovins, an American environmentalist and a founder of the Rocky Mountain Institute. See Waldman, "With Temperatures Rising, Here Comes 'Global Weirding.'" *New York Times* columnist Thomas Friedman also wrote about it. See, for example, Friedman, "The People We Have Been Waiting For," and Friedman, "Global Weirding Is Here." See Hayhoe, "What Is Global Weirding?" For an in-depth analysis of Hayhoe's Christian rhetoric, see Von Bergen and Mannon, "Talking Climate Faith." Special thanks to Lara Fowler, who first told me about Hayhoe and her work. Latour also uses the term "global weirding" (unattributed) in his 2011 lecture "Waiting for Gaia." Timothy Morton, too, discusses global weirding in *Dark Ecology*, 5–7.

4. For a helpful explanation of the atmospheric science behind the newly variable polar vortex, see Mitchell et al., "The Effect of Climate Change," 2608–18. The tweak is a brilliant stroke rhetorically, mainly because it closes the denialist loophole that exploits the misunderstanding about what global warming in fact refers to—an increase in the earth's core temperature, not (or not just) air temperatures. It is also the case, though, that because the phrase is designed for doubting ears and mouths, it could potentially strike people from other parts of the world, especially those sometimes referred to as the Global South, as too lighthearted or even flip.

5. *Oxford English Dictionary Online*, s.v., "weird, n.," Oxford University Press, accessed May 15, 2022, https://www.oed.com/view/Entry/226915.

6. Farrell, "The Weight of Rhetoric," 470.

7. Latour, "Why Has Critique Run out of Steam?," 231.

8. Lewis and Short, *A Latin Dictionary*, s.v., "intendo," 975. Michele Kennerly helpfully reminded me that intensity and attention inhabit similar etymologies.

9. Corrigan, *Black Feelings*, xxvii.

10. Corrigan, xvii.

11. Keller, *Recomposing Ecopoetics*. Elizabeth M. DeLoughrey pursues the awareness point, arguing that "our increasing awareness of climate change is catalyzing new imaginaries and, by extension, new allegorical forms to address the dynamics of our planet"—DeLoughrey, *Allegories of the Anthropocene*, 1. A non-exhaustive list of key publications about the term Anthropocene include, foremost, Crutzen and Stoermer, "The 'Anthropocene,'" 17; Crutzen, "The 'Anthropocene,'" 13–18; Steffen, Crutzen, and McNeill, "The Anthropocene," 614–21; Menely, "Anthropocene Air," 93–101; Johnson et al., "After the Anthropocene," 439–56; Zalasiewicz, Waters, and Head, "Anthropocene: Its Stratigraphic Basis," 289. Stacy Alaimo offers a marvelous reflection on the Anthropocene in chapter 6 of her book *Exposed*, 143–68.

12. Nixon, *Slow Violence*, 2. My colleague Jeffrey T. Nealon phrases what he takes to be the matter with characteristic bluntness: "To paraphrase an observation of Frederic Jameson's, right now we're far better equipped to imagine extinction—the end of the world—than to imagine an alternative to global neoliberal capitalism." Nealon, *Plant Theory*, 121.

13. Here it might be helpful to invoke Wayne Booth's term "rhetrickery" as a label for the nefarious, manipulative acts of deception that often travel under the moniker of rhetoric. With a couple of key exceptions, such as the insight about competing magnitudes that chapter 4's exploration offers, instances of denial do not concern me in this book, even though they concern me more generally and probably ought to concern us all! Booth, *The Rhetoric of Rhetoric*, 9. Bruno Latour has a good deal to say about denialism in *Facing Gaia*, as does Sterman in "Communicating Climate Change Risks," 811–26. From the deliberation world, John S. Dryzek and Alex Y. Lo take on the question of getting through to deniers in "Reason and Rhetoric in Climate Communication," 1–16. Amitav Ghosh makes perhaps the most shattering of observations when he writes: "it would be a mistake to assume that denialism within the Anglosphere is only a function of money and manipulation. There is an excess to denialist attitudes that suggests that the climate crisis threatens to unravel something deeper, without which large numbers of people would be at a loss to find meaning in their history and indeed their existence in the world"—Ghosh, *The Great Derangement*, 138.

14. In this way, the book pursues a version of sensuous, felt rhetoric laid out in my previous book, with one crucial and constitutive difference: that instead of thinking about rhetorical theory as it was elaborated in the past, I am looking, along with the artists, scholars, and activists working right now, toward a future, one that is radically uncertain. Hawhee, *Rhetoric in Tooth and Claw*.

15. *Oxford English Dictionary Online*, s.v., "unprecedented, adj.," accessed December 14, 2021, https://www.oed.com/view/Entry/216439.

16. Consider, as so many have, as so many must, the atrocities wrought by settler colonialism, genocide, and the Holocaust. As the political philosopher Maurizio Passerin d'Entrèves puts it, "Faced with the tragic events of the Holocaust and the Gulag, we can no longer go back to traditional concepts and values, so as to explain the unprecedented by means of precedents, or to understand the monstrous by means of the familiar"—Entrèves, *The Political Philosophy of Hannah Arendt*, 3–4.

17. DeLoughrey, *Allegories of the Anthropocene*, 2.

18. Likewise, it is important to take care in calls for widespread implementation of Indigenous methods of reforestation and the like. Kyle Whyte, Chris Caldwell, and Marie Schaefer, "Indigenous Lessons about Sustainability," 149–79.

19. Chapters 4 and 5 illuminate this point with respect to the pandemic and to settler colonialism specifically.

20. All told, the press release, technical summary, and full report use the word "unprecedented" 113 times (not counting bibliographic references). Intergovernmental Panel on Climate Change, *Climate Change 2021*. The Press Release, Technical Summary, and full report may be found here: https://www.ipcc.ch/report/ar6/wg1/.

21. Latour, *Facing Gaia*, 43–44; Facourt, "Le taux de CO2 dans l'air." Latour's parenthetical quotation of Facourt changed the period at the end of the French sentence to an exclamation point.

22. Kyle Whyte calls into question the very language of crisis, offering instead coordination as a better frame. Whyte, "Against Crisis Epistemology," 52–64. I discuss his point further in chapter 5. Special thanks to Elizabeth Mazzolini for telling me about the crisis epistemology piece, and to the members of my Spring 2021 Crisis Rhetoric graduate seminar for discussing it with me.

23. For a useful and productive rhetorical frame on uncertainty itself as a mode of invention, see Walsh and Walker, "Perspectives on Uncertainty for Technical Communication Scholars," 71–86; and Walker and Walsh, "'No One Yet Knows What the Ultimate Consequences May Be,'" 3–34. It strikes me that the COVID-19 pandemic may have drawn out or perhaps even brought more nuance to this account of uncertainty as a source of argumentative invention. For a bundle of work on the importance of framing for public engagement with science, see research by Nisbet and his colleagues. This work approaches the matter from a media communication rather than a rhetorical standpoint. The two perspectives are complementary, though, and the account offered in this book shows how rhetoric offers a nuanced interpretation of how frames give issues resonance, rather than just *that* they do so. Frames can work much more deeply than their metaphorical names suggest. Nisbet and Mooney, "Framing Science," 56; Nisbet, "Communicating Climate Change," 12–23; Nisbet and Scheufele, "What's Next for Science Communication?," 1767–78.

24. Of course the matter of different disciplinary manifestations of precedent is a fascinating one. For a consideration of the principle of *stare decisis*, built on case-based precedent, see Duxbury, *The Nature and Authority of Precedent*; Bhala, "The Power of the Past," 873–978. Chaim Perelman, too, offers a useful reflection on *stare decisis*, the doctrine of legal precedent. See his *Justice, Law, and Argument*, 37.

25. Aristotle, *Rhetoric*, 1.9.1368a38–40, trans. Kennedy, 82–83. Michelle Ballif builds from such Aristotelian observations an alternative approach to deliberation based on divination. Ballif, "Divining Rhetoric's Future," 157–71.

26. Bindé, "Toward an Ethics of the Future," 67.

27. Monbiot, *Heat: How to Stop the Planet from Burning*, 205.

28. Ghosh's exact quotation is as follows: "let us make no mistake: the climate crisis is also a crisis of culture and thus of the imagination"—Ghosh, *The Great Derangement*, 9. An author of literary fiction himself, Ghosh joins a cast of literary critics, most notable among them Ursula Heise and Lawrence Buell, and also to a great extent Keller, in fruitfully elaborating imagination by means of literature (Buell and Heise both use the phrase "the environmental imagination"). Buell, *The Environmental Imagination*; Heise, *Imagining Extinction*; Heise, *Sense of Place and Sense of Planet*. Rhetoric and literature of course share a disciplinary corner with poetics, as a host of scholars have demonstrated, most notably Booth, *The Rhetoric of Fiction*; Kennerly, *Editorial Bodies*; and Walker, *Rhetoric and Poetics in Antiquity*. Though the poetics of imagination allowed by these two disciplinary perspectives in the context of climate change could occupy a whole book, in chapter 2 I discuss, by means of chronotopes, the perspectives and generic pathways that rhetoric and literature each allow, and my discussion of poetic witnessing later in this introduction pulls in this direction as well. The historian of science and technology James Rodger Fleming

focuses on challenges to the imagination posed by technologies of climate control. Fleming, "Picturing Climate Control: Visualizing the Unimaginable," 345–62.

29. Picard, "An Opening to Imagine the Present." For a riveting case for telling time differently in the climate crisis, see Bastian, "Fatally Confused," 23–48.

30. Picard, "An Opening to Imagine the Present."

31. Medina, *The Epistemology of Resistance*, 7.

32. Whyte, "Time as Kinship," 39–55.

33. For a useful elaboration of how *kairos* entails *chronos*, see Detweiler and Pantelides, "Predicting Futures, Performing Feminisms," 157–70. I will return to their article in chapter 3.

34. By "bulk" I wish to suggest a necessary accumulation. Key scholars in rhetoric who write on witnessing include Michael Bernard-Donals, whose work on memory and rhetoric draws out memory's "'incantative' transmission—the question of how this nearly impossible thing, this void makes its way from the one who was there to those who were not and could not have been there" (Bernard-Donals, *Forgetful Memory*, 172.) See also Bernard-Donals, "The Rhetoric of Disaster," 73–94. My colleague Bradford Vivian, whose work I have already mentioned, provides insights that ground my own thinking on witnessing, as I make evident in the subsequent discussion. Barbie Zelizer, in *Remembering to Forget: Holocaust Memory Through the Camera's Eye*, brings her expertise on visual rhetoric and photography to bear on Holocaust memory and witnessing. Likewise, Justin Ewalt considers technologies of witnessing in the context of the Holocaust with his "Mapping Injustice: The World Is Witness," 333–54. In philosophy and literary and cultural studies, Kelly Oliver takes a psychoanalytic approach to witnessing, memory, and forgetting, an approach Michael Rothberg explicitly turns from to theorize witnessing in public-facing contexts such as the law, the media, and the like. Oliver, *Witnessing: Beyond Recognition*; Rothberg, "Between Auschwitz and Algeria," 162. Other key writings are Felman, "Fire in the Archive," 48–68; Levine, *The Belated Witness*; and Agamben, *Remnants of Auschwitz*. Jacques Derrida writes explicitly about a poetics of witnessing in *Sovereignties in Question*, 75–76.

35. I use the word "keeping" here deliberately. In doing so, I am thinking of arguments that identify witnessing as the defining act of our time, including that advanced by Annette Wieviorka in *The Era of the Witness*. In her examination of humanitarian witnessing, the political theorist Michal Givoni writes that "acts of witnessing and testimonial narratives have become so deeply integrated into political life that it is hard to imagine politics without witnesses and testimonies"—Givoni, *The Care of the Witness*, 4. John Ellis's account of television-mediated witnessing names the twentieth century "the century of witness"—Ellis, *Seeing Things*, 9. And Bradford Vivian, whose book title posits that witnessing has in fact become commonplace, presents "commonplace witnessing [as] a vital and pervasive mode of influence born in the crucible of modern public culture and intensified in its late modern, or contemporary permutations"—Vivian, *Commonplace Witnessing*, 2.

36. It is of course possible that these elements of witnessing have been there all along. But again, the point here is about intensification: they are being drawn out, assigned priority, made newly urgent by the changing climate.

37. Vivian, *Commonplace Witnessing*, 7. Elsewhere Vivian links witnessing's morality to time, observing that "witnessing rhetorically evokes and shapes collective perception of time so as to commend appropriate historical, political, and moral judgments"—Vivian, "Witnessing Time," 206. I will return to his arguments about temporality later in this introduction and in chapter 3. Giorgio Agamben finds that the movement toward "justice and truth" lend acts of witnessing a kind of "fullness"—Agamben, *Remnants of Auschwitz*, 34.

Megan Poole's framing of scientists as witnesses presents an open, relational account of witnessing, one that ties it to listening—Poole, "Witnessing the Open Semiosis."

38. Peters, *Courting the Abyss*, 245.

39. Fukushima, *Migrant Crossings*, 23.

40. Vivian, "Witnessing Time," 208. This ability of witnessing to slide through time informs Rothberg's important conception of "multidirectional memory." The time in this case is mainly past and present. Rothberg, "Between Auschwitz and Algeria," 158–84.

41. Sheikh, "The Future of the Witness," 147–48.

42. Foust and Murphy, "Revealing and Reframing Apocalyptic Tragedy," 152. Ismael Quiñones presents a theory of now and/as future by rethinking the very idea of texts. See Quiñones Valdivia, "Text as a Nowing," 141–59. His in-progress work on "undocumented time" will be worth the wait.

43. The toxic tours that Pezzullo studies provide, as she puts it, "opportunities to witness"; they bring before the senses what has happened, what is happening still. Pezzullo, *Toxic Tourism*, 9–10, 31, 145. Julie Kalil Schutten finds in Pezzullo's work "the rhetorical potential of witnessing" and urges its expansion. Schutten, "Environmental Sustainability," 339–42.

44. *Oxford English Dictionary*, s.v., "witness, v.," accessed December 23, 2021, https://www.oed.com/view/Entry/229714?rskey=BlWSR5&result=2&isAdvanced=false.

45. Peters, *Courting the Abyss*, 257.

46. The quoted phrase is from Peters, "Witnessing," 713.

47. In rhetoric, see Pezzullo, *Toxic Tourism*, 9, and two essays by Joshua Trey Barnett: "Vigilant Mourning and the Future of Earthly Coexistence," 13–33, and "Naming, Mourning, and the Work of Earthly Coexistence," 287–99. In art (and an article I will revisit in chapter 5), see Ross, "The Aesthetics of Coexistence as Ongoing," 169–82. In object-oriented philosophy, see Morton, *Dark Ecology*, 1.

48. Hawhee, *Rhetoric in Tooth and Claw*, 79.

49. Here I wish to call attention to, and to thank the editors of, the 2017 special issue of *Rhetoric Society Quarterly* whose "rhetorical bestiary" moves in this direction. They include a tantalizing mention of the bestiary as a potential mode of witness. See Gordon, Lind, and Kutnicki, "A Rhetorical Bestiary," 224.

50. Bastian, "Fatally Confused," 23–48.

51. Boswell, "Climates of Change," 9.

52. Grieve, "Environmental Justice Witnessing," 968–85. As Grieve phrases it, "in bare terms, environmental justice seeks justice for multiple species, not just humans while witnessing theorizes the way injustice is storied, remedied, memorialized, and worked through" (975–76).

53. Alaimo, *Bodily Natures*, 2.

54. *A Witness Tree* is the title of a 1942 collection of Robert Frost's lyric poems. See also Mapes, *Witness Tree*. The battlefield at Gettysburg, Pennsylvania has a number of trees marked as "witness trees" that are known to have been alive during the Battle of Gettysburg in 1863. Some have bullets from the battle lodged in their cores. For a very brief reflection on "wounded trees" as witness trees, see Lyons, "An Embodied Landscape," 62. A fascinating common use of the phrase "witness tree" can be found in forestry and ecological research as well. The label is often used without definition, but S. Andrew Predmore, Josh McDaniel, and John S. Kush offer helpful insight: "The terms witness tree and bearing tree are used interchangeably and refer to trees marked by land surveyors to identify parcels of land in preparation for sale." Witness trees were therefore a tool of settler colonialism, and they remain a marker of corners in lot boundaries. Predmore, McDaniel, and Kush. "Pre-

settlement Forests and Fire in Southern Alabama," 1723–24. See also Cogbill, Burk, and Motzkin, "The Forests of Presettlement New England, USA," 1279–1304. Trees will return later in this book as witnesses in chapter 5 and in the epilogue.

55. Martellozzo, "Wind, Wood, and the Entangled Life of Disasters," 432.

56. Cohen and Duckert, "Introduction: Eleven Principles of the Elements," 4.

57. Cohen and Duckert, 4.

58. Perelman and Olbrechts-Tyteca, *The New Rhetoric*, 118–20.

59. Perelman, *The New Rhetoric and the Humanities*, 17.

60. Vivian, "Witnessing Time," 213. This radical absence is exactly what leads Peters, rightly, to figure the witness as a medium. Peters, *Courting the Abyss*, 244.

61. Perelman, *The New Rhetoric and the Humanities*, 17.

62. Blair, "Contemporary US Memorial Sites," 18.

63. Farrell, "Sizing Things Up," 6.

64. Farrell, "The Weight of Rhetoric," 484.

65. Finnegan, *Making Photography Matter*, 3.

66. Olson, *American Magnitude*, 14.

67. The distinctly American version of magnitude, as Olson's final chapter shows, is perhaps something best moved past. Olson, *American Magnitude*, 180–94.

68. Rice, "The Rhetorical Aesthetics of More," 26.

69. Rice, 38.

70. The scale was developed at the Dartmouth Flood Observatory and is known as the DFO magnitude. The scientific articles that use the DFO magnitude are numerous, but for a detailed discussion, see Stephens, Clarke, and Nicholls "Trends in Reported Flooding," 53–55.

71. Schneider and Walsh, "The Politics of Zoom," 1–11. Timothy Morton's "hyperobjects" are almost unimaginable in scale—Morton, *Hyperobjects*.

72. Dobrin and Morey, "Mediating Interfaces," 1–14; Van Lieshout et al., "The Power to Frame the Scale?," 550–73; Reilly and Adamowski, "Spatial and Temporal Scale Framing." Maxwell Boykoff considers the problem of scale as well—Boykoff, *Creative (Climate) Communications*, 157.

73. O'Gorman's quick pivot may give the impression that he is conflating the two, but his leaning on comparison is what pulls magnitude into scale. O'Gorman, "Aristotle's *Phantasia* in the *Rhetoric*," 28–31. Whereas Olson finds the sublime in magnitude (Olson, *American Magnitude*, 16–18), Peters presents the sublime as a matter of human scale in a passage worth quoting: "Contrary to everyday usage, of course, *sublime* does not just mean glorious or wonderful. It means the suspension of the human scale. Maybe quantity does have something to do with humanity?" Peters, "33 + 1 Vignettes," 311.

74. For a helpful diagnosis of the rhetorical challenges posed by scale from ecological point(s) of view, see DiCaglio, Barlow, and Johnson, "Rhetorical Recommendations Built on Ecological Experience," 438–50. The literary critic Derek Woods focuses on scale in his work and even obliquely links it to the absence/presence discussed above when he examines how figurative language helps "make alien scales present in the collective life of politics." Woods, "Scale in Ecological Science Writing," 127. See also Woods, "Scale Critique for the Anthropocene," 133–42. Joshua DiCaglio's book-length consideration of scale offers helpful insights on scale and resolution as well as scale and mysticism. DiCaglio, *Scale Theory*, 23–24, 212–19.

75. DeLoughrey, *Allegories of the Anthropocene*, 2.

76. In a stunning account of early geometric forms, Jeanne Fahnestock documents the

definition of quantity itself (by Melanchthon) as a kind of magnitude. See Fahnestock, "The New Mathematical Arts of Argument," 178.

77. *Oxford English Dictionary*, s.v., "magnitude, n."

78. I am grateful to an anonymous reader whose skepticism helped me see and clarify this distinction between scale and magnitude.

79. Some may characterize my approach in this study as "neoclassical," but the point is that some ancient terms (such as *kairos*, which long predates the "classical") continue to circulate and are changing as they do so; indeed, they must change. See Carey, "Necessary Adjustments," 269–86. I return to this point, and to Carey's work, in chapter 3.

80. Richard Besel's work on James Hansen's Senate testimony is exemplary here, as is Daniel Brouwer's examination of ACT-UP members' congressional testimony, which I draw on in chapter 3. Besel, "Accommodating Climate Change Science," 137–52; Brouwer, "ACT-ing UP in Congressional Hearings," 87–109.

81. In *Rhetoric in Tooth and Claw*, for example, I followed nonhuman animals through some of rhetoric's oldest texts and traditions, and in the very different *Moving Bodies*, I followed the bodies through Burke's writings. Methodologically, those books did not require that I consult with my university's Institutional Research Board, as I did twice with this study.

82. Jenell Johnson elaborates a concept of "visceral publics" in her analysis of a mid-century public health controversy, and Emily Winderman, Robert Mejia, and Brandon Rogers combine Johnson's concept with sensory rhetoric to examine how smell constitutes public health matters. Stephanie Larson, whose work on magnitude I discuss further in chapters 2 and 4, has given the field a theory of visceral rhetoric, which she describes as "instances when the body responds, reacting to words or actions." Larson, *What It Feels Like*, 2; Johnson, "'A Man's Mouth Is His Castle,'" 1–20; Winderman, Mejia, and Rogers, "'All Smell Is Disease,'" 115–46. In anthropology, Kath Weston places the visceral, viscerality, and visceral sense at the center of her writing about the animation of eco-intimacies in a "socially manufactured, recursively constituted 'environment'"—Weston, *Animate Planet*, 8. Boykoff, who helpfully zeros in on the role of humor in climate change communication, issues a call for more research on what he calls "short- to long-term impacts of experiential, affective, visceral, tactile, aesthetic and emotional influences—as they relate to climate change engagement." Boykoff, *Creative (Climate) Communications*, 129, 111–27. Rhetoric scholars, as this book documents, are working on all of these fronts.

83. A good example of the kind of cognitive, behavioral approach I am not taking here is offered by Zaval and Cornwell in "Cognitive Biases, Nonrational Judgments, and Public Perceptions of Climate Change." For more on deliberation and emotion in the context of climate change, see Roald and Sangolt, *Deliberation, Rhetoric, and Emotion*.

84. A note on this work's relationship to existing scholarship in the environmental humanities, eco-philosophy, and climate communication: These are all rich and often themselves interdisciplinary areas of study. As will become clear to readers who spend time with this book, such work informs and enables my own, and I engage that work throughout the book. Even so, I tend to avoid a "corrective" approach to existing scholarship, because that approach can quickly become tedious at best, imperious at worst. So much has been written about the climate crisis, and so much more remains to be said. *A Sense of Urgency* aims to articulate, specifically, what rhetoric, as an intensifier, brings to the very large table, and how rhetorical concepts and practices are evolving before our eyes.

85. As Charles Bazerman puts it in a characteristically frank assessment of case-study approaches, "No study focuses on what remains the same, stabilizing tendencies, or cases

of 'nothing going on here'"—Bazerman, "Afterword: Social Changes in Science Communication," 270. For a useful discussion of case study methods, see Moriarty et al., "Durable Research, Portable Findings," 124–36.

86. The variety may well result in a somewhat sprawling book that may strike some scholars as unwieldy, or in need of a control of some sort. Yet this is not an empirical study—it is a theoretical one, along the lines of what Kirt Wilson labels "theory/criticism." (My thanks to an anonymous reviewer for characterizing this work in that way.) Wilson, "Theory/Criticism," 280–96. Furthermore, climate rhetoric is by its very nature unwieldy, and it takes (or ought to take) a variety of forms. It cannot and must not hold still. This book's range of departures honors that point.

87. Credit for the phrase "spheres of resonance" goes to St. Amant and Graham's "Research That Resonates," 104, 111, though they use it as a broader category to think about societal, institutional, and knowledge domains. Here, the knowledge domains of memory studies, public address, public health, and public art guide the literature I draw on to help illuminate salient concepts, though obviously each one contains multiple disciplinary perspectives. One anonymous reader opted to frame the chapter foci as "genres," but I find that word a little too static.

88. Timothy Morton's discussion of global warming as "a *wicked problem*—or even as a *super wicked problem*" is instructive here. Morton, *Dark Ecology*, 36–38.

Chapter Two

1. National Snow and Ice Data Center, "Dead Ice."

2. These downloadable images appear in NASA Earth Observatory's "Image of the Day" online feature. Hansen, "Okjökull Remembered."

3. For a geological overview of Ok and other glaciers in this region of Iceland, see Björnsson, "Glaciers of the Central Highlands," 275–336.

4. These are my translations with the help of Icelandic dictionaries: Icelandic for excessive summer heat—*Óhófleg sumarhlýindi*; "Nothing was done to save"—*Ekkert var gert til bjargar*. When he shared the death certificate with me via email, Sigurðsson was careful to note that the certificate is ceremonial and not legal. In pointing this out to me, Sigurðsson noted that an Argentinian glaciologist was arrested for his definition of glaciers. Oddur Sigurðsson, email message to Debra Hawhee, October 10, 2019. For details on the case Sigurðsson mentions in the email, see Tollefson and Rodríguez Mega, "Geoscientist Faces Criminal Charges," 159–60.

5. Howe, virtual interview.

6. Here I am invoking Kendall Phillips's positioning of rhetoric "as an art of crafting public sentiment," which I find eminently helpful for both its brevity and its emphasis on sentiment or feeling, which turns out to be the potent force of imaginative strategies this chapter illuminates. Phillips, "The Failure of Memory," 218.

7. Howe, quoted in a popular press story (Engel, "Scientists Unveil Memorial"). The subtitle of Howe and Boyer's 2018 documentary also carries this idea of small story: "A Little Movie about a Small Glacier at the End of the World."

8. Perhaps one feature of this "renovation" ought to be noted at the outset: my choice to translate *topos* as place rather than space, a distinction that has come to matter for theorists, especially those working in rhetoric and geography, but particularly scholars working on public memory in the context of rhetoric. As Bradford Vivian observes in his account of public memorial services held in the US on September 11, 2002, "The role of place holds inestimable significance in the enactment of these commemorative rituals" ("Neoliberal Epi-

deictic," 6). While *topos* holds both possibilities—it can mean a space left open in a document, for example (Liddell, Scott, and Jones, *Greek-English Lexicon*, s.v., "τόπος"), it more frequently carries valences of specific places or regions, and even—and this set of meanings is helpful for this chapter's connection to memorial studies—of significant public sites. For a comprehensive survey of the word's pre-Aristotelian and Aristotelian entailments, see Miller, "The Aristotelian *Topos*," 132–35, and Miller, "Aristotle's 'Special Topics,'" 61–70.

9. Julie Doyle's work on the aesthetics of comparative frames presented by time-interval photography, for example, depends on a representational model of communication, one that comments on "public perception" in a powerful way. Doyle, "Picturing the Clima(c)Tic," 133. The photographs examined by Doyle work like the NASA images of Okjökull presented at this chapter's outset, and Doyle helpfully documents their work in making visible evidence that is not immediately accessible (Doyle, 147).

10. Peters, "33 + 1 Vignettes," 305–31. I return to this work in chapter 4.

11. In research that shows the continued need to revitalize the concept of the chronotope, John Lynch also identifies a generational chronotope at work in the rhetoric of US presidents. For Lynch, "A generational chronotope distinguishes two broad groups of people, divided by time of birth"—Lynch, "Generational Chronotopes and Accounting for Unethical Medicine," 287. But whereas Lynch's case study presents a generational chronotope that divides generations for the purposes of assigning blame for past harm, the generational chronotope I follow in this chapter is a connective one, a strategic bond that tries to anticipate and forestall harm. Together, Lynch's essay and this chapter showcase once again the flexibility of chronotopes, even ones that warrant the same label.

12. Buell, *Writing for an Endangered World*, 6. Other scholars working in and around rhetoric, Phaedra C. Pezzullo foremost among them, have elaborated the important of care work to environmental communication. Pezzullo helpfully frames environmental communication as "a care discipline." Pezzullo, "Environment," 582; see also Pezzullo and de Onís, "Rethinking Rhetorical Field Methods on a Precarious Planet," 115; Pezzullo and Cox, *Environmental Communication and the Public Sphere*, 9–10. See also Barnett, "From Ecological Grief to *Gelassenheit*," 793–94. I will return to Barnett's essay in the epilogue. See also Barnett, *Mourning in the Anthropocene*. For a capacious, philosophical reflection on the ethics of care, see Bellacasa, *Matters of Care*.

13. Bakhtin, *Dialogic Imagination*, 250.

14. Bakhtin, 98.

15. Mandel, *Station Eleven*.

16. Bakhtin, *Dialogic Imagination*, 243, emphasis in original.

17. For the education researchers David Bloome and Laurie Katz, chronotopes are conceptions of space and time that frame school activities and literacy practices ("Literacy as Social Practice and Classroom Chronotopes," 222). But the framing work is important. The linguistic anthropologist Kristina Wirtz presents chronotopes as linguistic organizers of experience, thereby illuminating their world-making work. See Wirtz, "The Living, the Dead, and the Immanent," 343–69.

18. I am by no means the first theorist of rhetoric to find salience in Bakhtin's concept of chronotopes, though my emphasis on their ability to shape the quality of attention departs somewhat from existing work, which tends to use chronotopes most commonly as a tool for showing the ideological dimensions of conceptions of time, as in Jordynn Jack's work on public arguments about genetically modified foods or Catherine Schryer's on the way chronotopes work in scientific research articles to exert power and control. In 2010, Jodie Nicotra and Judith Totman Parrish, in a collaborative essay that deserves more consideration than it has received, use chronotopes to identify ideologies at work in global

warming discourse, particularly the "time compression" that works against, and prevails over, a knowledge of deep geological time. Deep time is something Parrish, as a paleoclimatologist, knows well. In their analysis, Nicotra and Parrish move briefly past the neoliberal ideological implications of time-space compression to acknowledge the affective work of time: "time acts as more than just a theoretical concept—it also has an affective dimension." Katherine Flowers, too, uses chronotopes as a particularly illuminating concept for examining public arguments about a town's officially adopting an "English-only" policy. And Laurie Gries, whose work features in chapter 4, grounds her new materialist conception of rhetorical transformation in a "chronotopic understanding of time," an understanding that acknowledges the chains (Prior and Shipka, cited below, call these "laminations") of concepts that crystallize anew with different iterations. See Gries, *Still Life with Rhetoric*, 42–45. Jack, "Chronotopes: Forms of Time in Rhetorical Argument," 52–73; Schryer, "Genre Time/Space," 81; Nicotra and Parrish, "Rushing the Cure," 230; Flowers, "Making English Local," 469–92.

19. Stewart, "Weak Theory in an Unfinished World," 77.

20. Bakhtin, *Dialogic Imagination*, 252.

21. Lemon, "Sympathy for the Weary State?," 845; Prior and Shipka, "Chronotopic Lamination," 180–238. Prior's book attunes chronotopes to writerly experience; there, with the aid of the chronotope, he finds what he usefully calls "affective atmospheres." Prior, *Writing/Disciplinarity*, 247.

22. In fact, the inability to consider timescales at once is evident in the somewhat panicked dismissal published alongside Nicotra and Parrish's piece by Marlia Banning, in which Banning worries that the deep time put forward by Nicotra and Parrish ignores human time, thereby almost proving Nicotra and Parrish's point about the difficulty of thinking beyond human time. Banning, "Exigency in Dispute," 638–60.

23. Henry Johnstone was the first to define rhetoric as "an art of getting attention," in "Rhetoric as a Wedge," 334. See also Oakley, *From Attention to Meaning*. Kristie Fleckenstein's work with Katherine Hayles's categories of deep attention and hyper-attention in her consideration of the stereoscope gets at one stratum of quality, and she usefully discusses attention in relation to materiality. Fleckenstein, "The Nineteenth-Century Parlor Stereoscope," 125–38. For more on care, see Tronto, *Moral Boundaries*, and *Caring Democracy*. In rhetoric, Phaedra Pezzullo and Catalina de Onís foreground the importance of care for environmental rhetoric, building anticipation for future work from Pezzullo on the matter. Pezzullo and de Onís, "Rethinking Rhetorical Field Methods." Joshua Trey Barnett, too, has drawn out the matter of care in his work on ecological grief—Barnett, "Grievable Water," 273–91, and *Mourning in the Anthropocene*.

24. The comparison to knots is Bakhtin's: "The chronotope is the place where the knots of the narrative are tied and untied" (*Dialogic Imagination*, 250).

25. Bakhtin, *Dialogic Imagination*, 248.

26. Lynch affirms the importance of this "feltness" of chronotopes, linking it to the urgency of *kairos* in the context of harm done to humans in the name of medical research. See Lynch, "Generational Chronotopes," 286.

27. Vivian, *Commonplace Witnessing*, 5.

28. As Howe put it to me in our conversation, "we're the experts in ceremony, and ritual, and human acts . . . and climate change is not an environmental problem, it's a human problem." And as the anthropologist Anand Pandian, in an interview alongside Howe, says of anthropology, "it is this discipline that has dedicated itself most doggedly to an investigation of the human, *anthropos*, as a problem and a horizon"—Picard, "An Opening to

Imagine the Present." And the epideictic genre is the perfect genre in which rhetoric and anthropology can meet—the genre, as the rhetoric scholar Laurent Pernot argues in his reflections on future directions of epideictic rhetoric, "invites an anthropological approach because of its ritual dimension." Pernot, *Epideictic Rhetoric*, 102. Indeed, as Cynthia Sheard notes, and as I elaborate elsewhere, the power of epideictic rhetoric inheres in its deliberative and imaginative functions, both of which are decidedly future-facing. Sheard emphasizes epideictic's "capacity to link thought with action, vision with reality, criticism with change." Sheard, "The Public Value of Epideictic Rhetoric," 788. I have written about vision and imagination in the context of epideictic rhetoric, as has Ned O'Gorman, in "Aristotle's *Phantasia* in the *Rhetoric*." Hawhee, "Looking into Aristotle's Eyes," 139–65. Pernot writes that epideictic's "chief function is a social one. It gives a shape to the representations and common beliefs of the group; it renders explicit, and justifies, accepted values, and on occasion it even offers lessons in new values." Pernot, *Epideictic Rhetoric*, x. Chaim Perelman and Lucie Olbrechts-Tyteca state plainly that the purpose of epideictic "is to increase the intensity of adherence to values held in common by the audience and the speaker." My analysis underscores their choice of the word "intensity" here. Perelman and Olbrechts-Tyteca, *The New Rhetoric*, 52. Accounts of the ideological functions of epideictic rhetoric are many, but central ones with expansive bibliographies may be found in Walker, *Rhetoric and Poetics in Antiquity*, and (again) O'Gorman, "Aristotle's *Phantasia* in the *Rhetoric*." Only a handful of scholars working on epideictic rhetoric consider the specific modes of argument within the genre. Laurent Pernot's is perhaps the most considered account and therefore guides my analysis.

29. Pernot, *Epideictic Rhetoric*, 87. For a crucial account of amplification in relation to energy, see Catalina M. de Onís's discussion in the introduction to her book *Energy Islands*, 10–12.

30. Pernot, *Epideictic Rhetoric*, 87–88.

31. Howe, virtual interview with author.

32. Howe, "Planned Remarks." It is important to note that because of the weather conditions, Howe shortened and extemporized her remarks. Even so, the script conveys something of the spirit and purpose of Howe's remarks.

33. Keller, *Recomposing Ecopoetics*, 2, emphasis in original.

34. Keller, 2. It is most definitely the case that environmental scientists and activists knew in the twentieth century about the damage being done to the planet; Keller's dating of the phenomenon to 2000 is meant to mark the broader spread of the awareness. See her justification on p. 8.

35. Pernot, *Epideictic Rhetoric*, 85.

36. "Acts of witnessing," Vivian argues, "are generically epideictic in nature." Vivian, "Epideictic Forgetting," 191.

37. Peters, "Witnessing," 722.

38. Bakhtin, *Dialogic Imagination*, 85; Blommaert, "Chronotopes, Scales, and Complexity," 110.

39. Johnson, "How to Mourn a Glacier."

40. Johnson, "How to Mourn a Glacier."

41. For a glimmering account of feminist hope for rhetorical theory, see Glenn, *Rhetorical Feminism*.

42. In the 1980s linguists noted a rise in both British and American English of certain forms of the subjunctive, including the were-subjunctive. Nichols, "The Suasive Subjunctive," 140–53. I should also note the optative mood (which does not exist in the English

language) as the mood of wish and desire—the plaintive and dramatic "if only . . . !" See Grosz, *On the Grammar of Optative Constructions*. John Durham Peters's arresting meditation on language in *Marvelous Clouds* entwines the optative and the subjunctive:

> Language has enlarged the circle of sociability among humans and stimulated not only human rationality but also our distinctive higher madness, such as the ability to recite epics, synthesize anthrax, and hunt whales; to call things into being that don't (yet) exist, such as wind tunnels, unicorns, and empires; to live in the future and the past, in the optative and the subjunctive modes; and to say "if " and "not." (Peters, *The Marvelous Clouds*, 261)

43. Aarts, "Subjunctive."

44. See, for example, the brilliant and careful work with the past-perfect subjunctive in Carole Blair, Balthrop, and Michel, "Mood of the Material," 6–20.

45. Zelizer, *About to Die*, 14–16.

46. Seligman, "Ritual and Sincerity," 17; Szerszynski, "Ecological Rites," 54.

47. This characterization comes in one of Geertz's many meditations on ceremony: "In a ritual, the world as lived and the world as imagined, fused under the same agency of a single set of symbolic forms, turns out to be the same world"—Geertz, *The Interpretation of Cultures*, 112.

48. Boyer, "Planned Remarks."

49. Barnett, "Grievable Water," 278. Barnett's lyrical work about the use of the photographic medium to make precarity visible, to transport feeling across distance, resonates with the work I am doing here, as does other work on ecological grief. For example, writing in 2010 on the matter of what makes something grievable, Catriona Mortimer-Sandilands notes that "there is lots of evidence of environmental loss, but few places in which to experience it *as* loss, to even begin to consider that the diminishment of life that surrounds us on a daily basis is something to be really *sad* about, and on a personal level"—Mortimer-Sandilands, "Melancholy Natures, Queer Ecologies," 338.

50. Cunsolo and Landman, "Introduction: To Mourn Beyond the Human," 3–4.

51. Howe and Boyer, unpublished proposal to Ragnar Frank, 1.

52. Howe, "Planned Remarks."

53. Before they started up the mountain, the prime minister read lines from the poem as well.

54. *Oxford English Dictionary*, s.v., "glacial, adj."

55. *Oxford English Dictionary*, s.v., "glacial, adj." The original source is Charles, "US Feels Heat on Global Warming Stance," 16.

56. National Snow and Ice Data Center, "Thermodynamics: Albedo"; *Oxford English Dictionary*, s.v., "albedo, n."

57. M. Carey, "The History of Ice," 497–527.

58. Carey, 500.

59. Sæþórsdóttir and Saarinen, "Challenges Due to Changing Ideas," 82–91; Welling, "Tourism, Landscapes, and Climate Change in Iceland." M. Jackson's *The Secret Lives of Glaciers* offers a useful, if not-so-"secret" cultural context for tourism vis-à-vis glaciers (196–222).

60. Howe, virtual interview.

61. Emphasis on long: "At the maximum of the last glaciation [about 22,000 years ago], Iceland was almost entirely covered by glaciers"—Einarsson and Albertsson, "The Glacial History of Iceland," 638.

62. This material is reported in Johnson, "How to Mourn."

63. "Assuming a rise of temperature of 2°C per decade, the glaciers of the central highlands will probably disappear within 100–150 years," writes the glaciologist Helgi Björnsson, in *The Glaciers of Iceland*, 275.

64. Bangerter and Lourwerse, "Focusing Attention with Deictic Gestures," 1331–36.

65. Prasch, "Toward a Rhetorical Theory of Deixis," 167.

66. Here is Buell on "thereness": "'thereness' is indispensable to the vision of place, whether virtually or immediately perceived . . . 'thereness' is elastic: places evolve, degrade, ameliorate." Buell, *Writing for an Endangered World*, 77.

67. Wirtz, *Performing Afro-Cuba*, 23; See also Wirtz, "The Living," 348, where she notes the deictic work of chronotopes in organizing experience.

68. Obama, "Remarks by the President at Exit Glacier."

69. Gore, "Remarks by Vice President Albert Gore," Glacier National Park. For an account of Gore's speech in the context of environmental history, Mark Carey's essay on the cultural history of glaciers begins with Gore's speech. Carey, "The History of Ice," 497–98.

70. Jakobsdóttir, "Iceland's Prime Minister: 'The Ice Is Leaving.'"

71. Jakobsdóttir.

72. Zemp et al., "Historically Unprecedented Global Glacier Decline," 74.

73. Lam and Tegelberg, "Witnessing Glaciers Melt," A05, 7–8.

74. Helgi Björnsson's comprehensive book on glaciers in Iceland explicitly presents mountain shapes as witnesses to "the erosive powers of glaciers"— Björnsson, *The Glaciers*, 4.

75. "The Last of Ice: Okjökull," *The Economist*.

76. Rice University News and Media News Release, "Lost Glacier to Be Honored with Memorial Monument."

77. Johnson, "How to Mourn."

78. Peters, "Witnessing," 716.

79. Air comprises up to 11 percent of glacial volume, according to a study of glacier density. Shumskiy, "Density of Glacier Ice," 568; Björnsson, *The Glaciers*, 3–5.

80. The "bear witness" phrase may be found in Björnsson, *The Glaciers*, 4.

81. Hmiel et al., "Preindustrial 14CH4," 409–12.

82. Zemp et al., "Historically Unprecedented Global Glacier Decline," 757.

83. Magnason, "The Glaciers of Iceland Seemed Eternal."

84. Zekollari, Huss, and Farinotti, "Modelling the Future Evolution of Glaciers," 1125–46.

85. Phillips and Reyes, "Introduction. Surveying Global Memoryscapes," 9–14.

86. Magnason, "The Glaciers of Iceland Seemed Eternal."

87. Howe and Boyer, *Not OK*.

88. This particular version appears in a podcast interview with Andrew Zuckerman, "Andri Snær Magnason." Other versions of the exercise can be found in Magnason's *Guardian* piece ("The Glaciers of Iceland Seemed Eternal"). In Howe and Boyer's *Not OK* documentary, Magnason formulates the exercise in the first person: "If my grandchild will become as old as my grandmother, the person that I will know and love the most in my life is still alive in the year 2150. I still have my daughter do this calculation I asked her, when is somebody that you love still alive? She did the math. 2170. 2170 is a person that she would know personally and love personally still alive. But we don't relate to our interaction to nature, resources, profit, corporations, politics, we don't relate in the real timescale. Not even in the timescale that actually belongs to our inner circle."

89. Magnason, "The Glaciers of Iceland Seemed Eternal."

90. Crick, "Conquering Our Imagination," 24, 28.

91. Buell, *Writing for an Endangered World*.

92. Nicotra and Parrish, "Rushing the Cure," 232.

93. Nicotra and Parrish, 230.

94. Thomas Farrell calls contingency "the habitat of rhetoric"—Farrell, "The Weight of Rhetoric," 472.

95. Howe, virtual interview.

96. Howe, "Planned Remarks." As Howe indicates, the lesson of being good ancestors is embedded in Indigenous lifeways, to use a word offered by Daniel R. Wildcat, a Yuchi member of the Muscogee Nation of Oklahoma, in his book *Red Alert! Saving the Planet with Indigenous Knowledge*. There Wildcat writes: "as my Haudenosaunee friends constantly remind us, we have a responsibility to live respectfully for our children seven generations into the future." Indigenous knowledge, knowledge that spans generations, is the kind of awareness that Howe, like Wildcat, believes is needed. Wildcat, *Red Alert!* 14–15. I want to be clear that I mean to quote Wildcat not to appropriate but to amplify, and with the same spirit of humility in which Howe presents Indigenous (North American) knowledges in the context of the memorial on Ok.

97. Howe, virtual interview.

98. Hawhee, *Rhetoric in Tooth and Claw*, 4, 13–15.

99. Cohen, *Stone: An Ecology of the Inhuman*, 16.

100. Cohen, 11.

101. Cohen, 17.

102. Howe, virtual interview. It is also the case that basalt rock has been found to be ideal for carbon sequestration—it is porous, and ocean basalt has acidic properties that can neutralize carbon.

103. Magnason, "A Letter to the Future."

104. Bakhtin, *Dialogic Imagination*, 245, emphasis in original.

105. As Eric Holthaus tweeted: "This is the first time in human history our planet's atmosphere has had more than 415 ppm CO_2. Not just in recorded history, not just since the invention of agriculture 10,0000 years ago. Since before modern humans existed millions of years ago. We don't know a planet like this." Holthaus, Twitter post.

106. Bakhtin discusses readers as chronotopes "in the completely real-life time-space where the work resonates"—*Dialogic Imagination*, 252.

107. O'Dowd, "Researchers Memorialize First Major Icelandic Glacier."

108. Bakhtin, *Dialogic Imagination*, 84.

109. This plaque also hints at the kind of anticipatory mourning that Joshua Trey Barnett views as necessary for "the work of earthly coexistence." Barnett, "Vigilant Mourning," 14.

110. Carole Blair was the first to argue that memorials "summon our attention to their assiduous materiality" (Blair, "Contemporary US Memorial Sites," 17). I expand upon this same quotation in chapter 5. Subsequent work, including work by Blair and her collaborators, has expanded that statement to show how attention is also summoned *through* that very materiality. See, in particular, two articles Blair co-wrote: Blair, William Balthrop, and Michel, "Mood of the Material," and Balthrop, Blair, and Michel, "The Presence of the Present." Two other key essays include: Blair, Jeppeson, and Pucci, "Public Memorializing in Postmodernity," and Blair, Dickinson, and Ott, "Introduction: Rhetoric/Memory/Place." See also Doss, *Memorial Mania*.

111. Howe and Boyer, unpublished proposal to Ragnar Frank, 1.

112. Stephanie R. Larson, "'Just Let This Sink in,'" 442.

113. The words in quotations are, again, from the second part of their documentary's title.

114. Johnson, "How to Mourn."

Chapter Three

1. Sengupta, "Greta Thunberg, on Tour in America." Franklin Foer uses the phrase "unvarnished tones" in "Greta Thunberg Is Right to Panic." The documentary film *I Am Greta* was released in the fall of 2020.

2. Felsenthal, "Person of the Year: The Choice."

3. Obama, "Just 16, Greta Thunberg Is Already One of Our Planet's Greatest Advocates."

4. Matthew Houdek and Kendall Phillips in fact use an excerpt from Thunberg's speech before the UN to open their introduction to the special issue of *Women's Studies in Communication* focused on rhetoric and temporality. There, Houdek and Phillips find in Thunberg's words "a distinct rhetorical emphasis on the now of the present moment." Houdek and Phillips, "Rhetoric and the Temporal Turn," 369. Thorough accounts of Thunberg's activities in North America may be found in these two news articles: Rice, "Greta Thunberg Is Sailing Across the Atlantic," and Sengupta, "Greta Thunberg Sets Sail, Again."

5. US Congress, House of Representatives, Subcommittee on Europe, Eurasia, Energy, and the Environment of the Committee on Foreign Affairs, "Voices Leading the Next Generation on the Global Climate Crisis," 116th Cong., 1st sess., September 19, 2019, 10.

6. Representative headlines include: "Greta Thunberg Has a Suggestion for How to Take Real Climate Action" (Asmelash, CNN News); "'Listen to the Scientists': Greta Thunberg Urges Congress to Take Action" (Milman and Smith, *The Guardian*); and "'I Want You to Take Action': Greta Thunberg, Teenage Climate Activist, Testifies Before Congress" (Hayes, *USA Today*).

7. Morton, *Being Ecological*, xviii–xx.

8. Sengupta, "Greta Thunberg, on Tour in America."

9. More accounts of Thunberg's rhetoric are, I'm sure, in the works. An early instance, and one that examines the UN Climate Action Summit speech mentioned below, is presented in Oltof, "Projecting a Future Present," 67–82. Media analyses, which are different from but still complementary to rhetorical examinations, include Murphy, "Speaking for the Youth, Speaking for the Planet," 193–206; and Ryalls and Mazzarella, "'Famous, Beloved, Reviled, Respected, Feared, Celebrated,'" 438–53.

10. The media obsession with Thunberg at the expense of the other youth activists around the world marks another potential drawback to what Maxwell T. Boykoff and Michael K. Goodman call "the 'celebritization' of climate change." It should be noted that their work predates Thunberg's rise to celebrity status *as* a climate activist and therefore focuses mainly on celebrities who bring fame established in other areas (the entertainment industry, for example) to the matter of climate change. See Boykoff and Goodman, "Conspicuous Redemption?," 395–406. Julie Doyle, Nathan Farrell, and Michael K. Goodman dive even more deeply into the matter of this celebritization. They find that the resulting spectacle "may render climate change as monetized and meaningless media performances rather than interventions related to the intricate social practices of the everyday or political realms." See Doyle, Farrell, and Goodman, "Celebrities and Climate Change," 96. For an expansive consideration of media and climate, see Boykoff, *Who Speaks for the Climate?*

11. In fact, the fiery remarks Thunberg would deliver four days later at the UN Climate Action Summit were much more of a piece with the rhetoric presented by her American

peers at these two congressional hearings. There, Thunberg began, "My message is that we'll be watching you. This is all wrong. I shouldn't be up here. Yet you all come to us young people for hope. How dare you." Thunberg, "Transcript: Greta Thunberg's Speech at the UN Climate Action Summit."

12. Our Children's Trust, a nonprofit public interest law firm, led a protracted federal suit against the US government and has initiated legal proceedings in each of the fifty states. See Juliana v. United States, 18-36082 (9th Cir. 2020) and Our Children's Trust, "Legal Proceedings in All 50 States."

13. Alter, Haynes, and Worland, "Person of the Year: The Conscience," 55.

14. "Rules of the One Hundred Sixteenth Congress," H. Res. 6 (January 9, 2019) at 43–44.

15. The original goal of the Select Committee was to make policy recommendations by March 2020, but the COVID-19 pandemic extended that timeline. The majority members of the Select Committee released a 547-page report in June 2020, drawing on the seventeen official hearings and six member-level roundtables or discussions held by the Select Committee to that point. Select Committee on the Climate Crisis, "Solving the Climate Crisis: The Congressional Action Plan for a Clean Energy Economy and a Healthy, Resilient, and Just America," June 2020, 17–18, https://climatecrisis.house.gov/sites/climatecrisis.house .gov/files/Climate%20Crisis%20Action%20Plan.pdf.

16. Subcommittee on Europe, Eurasia, Energy, and the Environment, "Voices Leading," 2.

17. Galloway and United States Congress, *History of the United States House of Representatives*, 80.

18. Brouwer, "ACT-ing UP in Congressional Hearings," 88.

19. Barrett, "Yes, I'm Striking Over the Climate Crisis."

20. Brouwer, "ACT-ing UP in Congressional Hearings," 92. Michael Halloran's early essay on "doing public business in public" helpfully identifies a genre of "public proceeding" whereby a body of leaders conducts its official business. As an informational hearing, this hearing explicitly does not quite count as what Halloran labels "public proceeding," which has "a decision to be made," but it does share an important feature of those kinds of proceedings, where audiences, composed of constituents, are witnesses who "'overhear' the arguments not addressed to them." Halloran, "Doing Public Business in Public," 119, 120. Brouwer identifies something similar here with the work of witnessing. Thanks to Bradford Vivian for recommending Halloran's piece to me.

21. US Congress, House of Representatives, Select Committee on the Climate Crisis, "Generation Climate: Young Leaders Urge Climate Action Now," 116th Cong., 1st sess. (201) at 41.

22. Subcommittee on Europe, Eurasia, Energy, and the Environment, "Voices Leading," 3.

23. *Oxford English Dictionary*, s.v., "galvanize," "galvanic," and "galvanism."

24. To be sure, many read Thunberg's testimony as a cause for celebration, but I am more inclined to read it as a missed opportunity. One example: Dr. Lucky Tran, "'I have not come with any prepared remarks, I am instead submitting the IPCC report as my testimony because I don't want you to listen to me. I want you to listen to the scientists.' @GretaThunberg With the A+ mic drop in front of Congress!" @luckytran, Twitter post, September 16, 2019, https://twitter.com/luckytran/status/1174413101413781504?lang= bn. The possible explanations for Thunberg's choice are many and should be listed here for the sake of clarity. Thunberg's testimony was most likely meant, as she suggests, to amplify scientific knowledge. Perhaps she wished to send a strong signal that she herself is done

arguing, that science should prevail, and that time is too short to make a case that it should. Perhaps it was an understandable overcorrection owing to her caustic public exchanges with the leader of this country's executive branch and the nation's inaction on climate change, which may have led her to believe that action would follow an understanding of the latest scientific findings. Donald Trump's swipes at Thunberg, confined mainly to Twitter, were covered widely. See, e.g., Taylor, "Trump Mocks Greta Thunberg on Twitter." Or her choice in this situation may have been a manifestation of her "superpower," the word she once famously used to describe her autism as that which sets her apart as a laser-focused spokesperson, and which she herself claims has led to difficulty she has had imagining motives for inaction apart from lack of access to or understanding of the available information. That superpower, which Erin Manning figures as autism's insurgency of neurodiversity, could well be behind the testimony. Manning, *The Minor Gesture*, 5. See Greta Thunberg, "When haters go after your looks and differences, it means they have nowhere left to go," @gretathunberg, Twitter post, August 31, 2019, https://twitter.com/gretathunberg/status /1167916177927991296.

25. See Bessette, *The Mild Voice of Reason*, 47. George B. Galloway writes that the primary function of congressional committee hearings "has been to collect facts so as to enable committee members to make informed judgments regarding legislative proposals" (*History of the United States House*, 81). Kristine Davis's analysis of congressional hearings finds that committee members rarely come to hearings to gather new information but rather to *verify* information (Davis, "A Description and Analysis of the Legislative Committee Hearing," 88–106). Tarla Rai Peterson, approaching legislative hearings from the point of view of rhetoric, notes that "a hearing provides an opportunity for groups to frame a situation around their interests" (Peterson, "The Rhetorical Construction of Institutional Authority," 261).

26. Tsing, *The Mushroom at the End of the World*, 20.

27. Gronbeck, "Rhetorical Timing in Public Communication," 84.

28. See, e.g., Corrigan, *Black Feelings*, discussed in more detail in the Introduction and below. Christa Olson, too, elaborates the importance of feeling in "American Magnitude," 17, 19–24.

29. Houdek and Phillips, "Rhetoric and the Temporal Turn." The other issue is *Rhetoric Society Quarterly* 51, no. 3 (2021), edited by Freda Buhre and Collin Bjork.

30. Poulakos, "Toward a Sophistic Definition of Rhetoric," 39.

31. The classicist John E. Smith was the first to discuss *kairos* as time's quality. See his "Time, Times, and the 'Right Time,'" 1–13; as well as his "Time and Qualitative Time," 46–57. For an elaboration of *kairos* as "rhetoric's time," see my "Kairotic Encounters," 18, and as a bodily sense of time, see my *Bodily Arts*, 65–85. Gronbeck, too, was largely talking about a "sense of timing," in line with the "fitting" or "appropriate" cluster of meanings packed into the Greek concept of timing, *kairos*, by now well worn in the scholarship. The more recent line of scholarship discussed below is rightly and subtly resistant to that set of meanings.

32. Chirindo et al., "Coda: A Rupture in Time," 459–70. Indeed, a perhaps incongruous pairing of this Coda with Gronbeck's 1974 article shines a light—a bright one—on the specific variables at work in the concept of timing, to use Gronbeck's terminology. Privilege and precarity, as the authors of the coda suggest, are exactly such variables.

33. For a brilliant qualitative study of temporality and precarity in the context of labor, see Sharma, *In the Meantime*.

34. Carey, "Necessary Adjustments," 269–86.

35. Carey, 270.

36. Carey, 270.

37. One exception worth noting on the *chronos-kairos* front is Detweiler and Pantelides, "Predicting Futures, Performing Feminisms," 157–70. Detweiler and Pantelides point out that "kairos would not be kairos without constraints, which often come in the form of chronos" (159). Even more tantalizingly and aptly, they write, "Chronos and kairos don't shoot in straight, parallel lines, they wind around each other" (168).

38. Carey, "Necessary Adjustments," 271.

39. Berlant, "Trauma and Ineloquence," 44.

40. Hunt, "The Rhetorical Mood of World War II," 1; Joshua Gunn rightly mentions Hunt's "elegant essay" as a possible precursor to the current interest in feeling and sensation, in Gunn, "Speech's Sanatorium," 29n15.

41. Hunt, "The Rhetorical Mood of World War II."

42. Charland, "Farrell's Moods," 350.

43. Corrigan, *Black Feelings*, xiv. Corrigan's book provides an outstanding model for studying rhetorical moods, even if she does not use that exact phrase.

44. Aristotle, *Rhetoric*, 2.1.1378a8, trans. Kennedy, 113.

45. Aristotle, *Rhetoric*, 2.1.1378a9, trans. Kennedy, 113.

46. Congressional Research Service, "Membership of the 116th Congress: A Profile," R45583, updated December 17, 2020, https://fas.org/sgp/crs/misc/R45583.pdf.

47. Select Committee on the Climate Crisis, "Generation Climate," 1.

48. "Generation Climate," 1–2.

49. "Generation Climate," 2.

50. "Generation Climate," 2.

51. "Generation Climate," 2.

52. As Houdek and Phillips observe, such a depiction of the present as "the moment of action" is "particularly common as a means of motivating action"—Houdek and Phillips, "Rhetoric and the Temporal Turn," 2.

53. Select Committee on the Climate Crisis, "Generation Climate," 2.

54. Subcommittee on Europe, Eurasia, Energy, and the Environment, "Voices Leading," 52.

55. "Voices Leading," 52.

56. Thunberg's other three sentences were these: "It is just such a thing that we should be taking for granted that we listen to the current best available united science. It is just something that everyone should do. This is not political opinions, political views, or my opinions, this is the science." Subcommittee on Europe, Eurasia, Energy, and the Environment, "Voices Leading," 52.

57. Select Committee on the Climate Crisis, "Generation Climate," 39.

58. US Census Bureau, "QuickFacts: Kinston city, North Carolina," https://www.census.gov/quickfacts/fact/table/kinstoncitynorthcarolina,lenoircountynorthcarolina/PST045219.

59. Select Committee on the Climate Crisis, "Generation Climate," 7.

60. "Generation Climate," 7.

61. "Generation Climate," 7.

62. "Generation Climate," 8.

63. "Generation Climate," 8.

64. "Generation Climate," 10.

65. Aristotle, for example, defines fear as "a sort of pain and agitation derived from the *phantasia* of a future destructive or painful evil" (*Rhetoric*, 2.5.1382a, translation modified from Kennedy, 128). See also Pfau, "Who's Afraid of Fear Appeals?," 216–37; and Nussbaum, *The Monarchy of Fear*.

66. See *Oxford English Dictionary*, s.v., "hesitation."

67. Select Committee on the Climate Crisis, "Generation Climate," 10.

68. "Generation Climate," 10. See also Select Committee on the Climate Crisis, "Solving the Climate Crisis," 19. Melody Zhang is also quoted in the report in a discussion of faith-based approaches to climate crisis (23).

69. Select Committee on the Climate Crisis, "Generation Climate," 10.

70. "Generation Climate," 17.

71. "Generation Climate," 34.

72. "Generation Climate," 15.

73. "Generation Climate," 35.

74. For a book-length meditation on the "as-if" in philosophical thought, see Appiah, *As If*.

75. Subcommittee on Europe, Eurasia, Energy, and the Environment, "Voices Leading," 18.

76. "Voices Leading," 18.

77. "Voices Leading," 18.

78. "Voices Leading," 18.

79. "Voices Leading," 19.

80. "Voices Leading," 12.

81. "Voices Leading," 12.

82. "Voices Leading," 12.

83. "Voices Leading," 12.

84. Margolin, *Youth to Power*, 88.

85. Margolin, 91.

86. Subcommittee on Europe, Eurasia, Energy, and the Environment, "Voices Leading," 12.

87. Margolin, *Youth to Power*, 88.

88. Margolin, 88.

89. Subcommittee on Europe, Eurasia, Energy, and the Environment, "Voices Leading," 12.

90. Margolin, *Youth to Power*, 41.

91. Margolin, 92.

92. Subcommittee on Europe, Eurasia, Energy, and the Environment, "Voices Leading," 13.

93. "Voices Leading," 13.

94. There are notable similarities between what Margolin does here and what the rhetorical critic G. Thomas Goodnight says the mid-century environmental activist and scientist Rachel Carson did: "contrasting temporal scales—the long-term redemption of the environment and the quickening pace of its destruction." See Goodnight, "Public Discourse," 430.

95. Right after Rep. Keating mentions that they are speaking to the future, he says, "However, let us be clear, we are also speaking to the present." Subcommittee on Europe, Eurasia, Energy, and the Environment "Voices Leading," 2.

96. "Voices Leading," 13.

97. "Voices Leading," 52. I have added emphasis based on Margolin's own emphasis evident in the recorded hearing.

98. "Voices Leading," 52–53.

99. Martin Luther King, Jr., "I Have a Dream."

100. Select Committee on the Climate Crisis, "Generation Climate," 34.

101. Subcommittee on Europe, Eurasia, Energy, and the Environment, "Voices Leading," 54.

102. "Voices Leading," 55.

103. "Voices Leading," 55.

104. "Voices Leading," 55–56.

105. Walsh, *Scientists as Prophets*.

106. During his presidency, Donald Trump made no secret of this hostility toward climate science, or toward Thunberg for that matter. See Jotzo, Depledge, and Winkler, "US and International Climate Policy under President Trump," 813–17.

107. "Word of the Year 2019," *Oxford Languages*, https://languages.oup.com/word-of-the-year/2019/.

Chapter Four

1. O'Hagan, "Keepsakes," 31.

2. Onder, Rezza, and Brusaferro, "Case-Fatality Rate and Characteristics of Patients."

3. Holshue et al., "First Case of 2019 Novel Coronavirus in the United States," 929–36. On California: Fuller and Baker, "Coronavirus Death in California."

4. In fact, the phrase "navigating the unknown" showed up in a pop-up ad to advertise a 99-cent subscription to *News Journal Online* when I was writing these very words and paused to look for news stories about journalists working in the pandemic.

5. Le Quéré et al., "Temporary Reduction in Daily Global CO_2 Emissions," 647–53.

6. Pongsiri et al., "Biodiversity Loss Affects Global Disease Ecology," 945–54; Mackenzie and Jeggo, "Reservoirs and Vectors of Emerging Viruses," 170–79; Estrada-Peña et al., "Effects of Environmental Change," 205–14.

7. Ye et al., "Zoonotic Origins of Human Coronaviruses," 1686–97; Mackenzie and Smith, "COVID-19: A Novel Zoonotic Disease," MA20013–MA20013; Contini et al., "The Novel Zoonotic COVID-19 Pandemic," 255–64. In addition to the direct relationship between the COVID-19 crisis and the climate crisis, experts have discussed numerous parallels between the two. For example, the pandemic disproportionately affected vulnerable communities and populations around the world, much like climate change does and will, as the youth rhetors discussed in chapter 3 made clear. Michael E. Mann's *The New Climate War* includes a wide-ranging discussion of parallels between COVID-19 and climate change, and lessons that can be taken from the former for the latter. I recommend the discussion in its entirety. Mann, *The New Climate War*, 238–51.

8. Markel, "What History Revealed." This piece draws from research Markel and his team conducted. Markel et al., "Nonpharmaceutical Interventions Implemented by US Cities," 644–54.

9. The phrase "flatten the curve" was included in a number of end-of-year reflections on new vocabulary 2020 taught us. Herrera, "Left Speechless? Let's Review," 1, 12; Wong, "13 Words or Phrases"; Broach, "17 Words, Phrases We Learned in 2020"; Dorman, "The Year 2020 in Words"; Netburn, "From COVID to Curbside"; McGreevy, "Covid-19, Cocooning, New Normal." The widespread recognition of the curve is yet another confirmation of an opening assertion made by James Wynn and G. Mitchell Reyes: "mathematics has always influenced public culture" (Wynn and Reyes, "Introduction," 1).

10. For example, I am adding this particular note on the second day of 2022 in the midst of the giant omicron wave, which is the sixth curve on the still-lengthening COVID-19 timeline. See Lukpat and Cahalan, "The US Breaks Its Single-Day Case Record."

11. Jack, "A Pedagogy of Sight," 192–209.

12. The phrase "general sense of the outbreak's magnitude" is third on a bulleted list of what an epi curve can reveal about an outbreak, according to the CDC's "quick-learn lesson" module. Centers for Disease Control and Prevention, "What Is an Epi Curve?"

13. Patricia L. Dunmire shows in detail how for Aristotle, in her words, "the future is understood to be a terminus of deliberation, rather than a means of persuasion in deliberative rhetoric." Dunmire, "The Rhetoric of Temporality," 85.

14. The explanatory power of G. Mitchell Reyes's theory of mathematics as a translative rhetoric takes concrete form in the story of the epi curve's temporary merger with the climate curves that this chapter follows. For Reyes, a rhetorical perspective provides clarity on mathematical relations to the world. Reyes, *The Evolution of Mathematics*, 11–12.

15. Perhaps the most well-known work in economics on this matter is early work on the idea of "availability" by Amos Tversky and Daniel Kahneman. Theirs is a cognitive approach that studies how judgments about probability and statistics are often based on "the ease with which relevant instances come to mind." Tversky and Kahneman, "Availability," 207. An excellent overview of this work and its implications may be found in Kahneman, *Thinking, Fast and Slow*, 50–70. The implications of their conclusions are massive, of course—Nobel Prize-winning level!—and I thank Mitch Reyes for pointing out the connection. An understanding of rhetoric as an art of intensification helps to think about how such ease of availability—such accessibility—operates on the level of feeling and, in this instance, by means of recognizable visual artifacts. Sensation and cognition work together to create the possibility for change.

16. See, for example, Schneider, "Image Politics," 191–209; O'Neill and Nicholson-Cole, "'Fear Won't Do It,'" 355–79; Sheppard, "Landscape Visualisation and Climate Change," 637–54; Schroth et al., "Visual Climate Change Communication," 413–32. Chapter 9 in Lynda Walsh's book presents graphics that are based on computer modeling as a "technology of divination" and provides a helpful overview of the challenges the IPCC faces with the rhetorical problems of clarifying uncertainty. Walsh, *Scientists as Prophets*, 176–77.

17. Walsh, "The Visual Rhetoric of Climate Change," 363–64. She argues that the semiotic approach to visual rhetoric would benefit from attention to *kairos*, a classical conception of time that I discuss in the introduction and in chapter 3 on felt time.

18. At least as she describes it: "This work has proceeded from one of three starting positions: extending the principles of Classical rhetoric to make them apply to visual as well as verbal symbols; working 'up' from the semiotics of visual symbols to their rhetoric; or, working 'down' from critical and cultural theory to detail how graphics distribute power in a polity." Each of the instances, as Walsh presents it, takes an "application" approach. Walsh, "The Visual Rhetoric of Climate Change," 262.

19. Gries, *Still Life with Rhetoric*, 88.

20. Finnegan, *Making Photography Matter*, 3, emphasis in original.

21. In this way, the graphic and its mantra filled some of the "narrative space" that crisis communication experts Matthew W. Seeger and Timothy L. Sellnow say is created by crisis: "Crisis creates a narrative space, a communication vacuum, or a meaning deficit that will be filled by stories told by those who experienced the crisis, crisis managers, journalists, and observers." In the case of the COVID-19 pandemic, I would say public health experts could be added to that list, and with them, their key data visualizations—one of their main means of communicating. Seeger and Sellnow, *Narratives of Crisis*, 13.

22. Hariman and Lucaites's observation is made in the context of iconic photographs, which for them "provide a particularly apt genre for exploring the relationship between a rhetorical artifact and audience response." Hariman and Lucaites, "Icons, Appropriations,

and the Co-Production of Meaning," 286. Laurie Gries's *Still Life with Rhetoric* pairs nicely with Hariman and Lucaites's argument because it adds insight in the context of digital circulation and virality.

23. As such, this chapter adds to the exciting examples of transdisciplinary relation between rhetoric and mathematics included in James Wynn and G. Mitchell Reyes's edited volume *Arguing with Numbers.* See also Reyes, "Stranger Relations." Where I land here also richly illustrates a point made by Reyes in the context of yet another crisis, the 2008 economic downturn: "mathematical discourse is a powerful system of translation and mediation out of which new hybrids emerge, and those hybrids can, under the right circumstances, 'transform the relations' of the networks that compose our worlds." Reyes, "The Horizons of Judgment in Mathematical Discourse," 91.

24. Ardern, "Prime Minister Jacinda Ardern Speaks After Coronavirus Committee Meeting."

25. Ardern.

26. Ardern.

27. Ardern.

28. Drew Harris posted his version of the curves with the all-important dotted line in February 2020. Harris, "Important to remember that #Covid-19 epidemic control measures may only delay cases, not prevent. However, this helps limit surge and gives hospitals time to prepare and manage. It's the difference between finding an ICU bed & ventilator or being treated in the parking lot tent," @drewaharris, Twitter post, February 28, 2020, https://twitter.com/drewaharris/status/1233267475036372992?lang=en.

Wiles and Morris credit Drew Harris and (now) the CDC as sources.

29. Roberts, "Flattening the Coronavirus Curve."

30. Harris, "Important to remember . . ."

31. Centers for Disease Control and Prevention, "Interim Pre-Pandemic Planning Guidance," 18.

32. Centers for Disease Control and Prevention, "Interim Pre-Pandemic Planning Guidance."

33. Siouxsie Wiles had this to say about the confusion over the graph's conceptual (generic) nature: "It's funny there are a lot of people criticizing it, saying . . . 'your areas aren't scientifically correct!' . . . The scientist bit of me is like, 'I know it's not perfect, but that's on purpose!'" Wiles, quoted in Wilson, "The Story Behind 'Flatten the Curve.'"

34. An interim version of the CDC graphic, designed by Rosamond Pearce in a story for *The Economist,* prompted Harris to redraw and share his graphic. See Roberts, "Flattening the Coronavirus Curve."

35. Edward Tufte recommends placing labels directly on the graphic—Tufte, *The Visual Display of Quantitative Information,* 87.

36. The .gif version may be viewed or downloaded at Wikimedia Commons: https://commons.wikimedia.org/wiki/File:Covid-19-curves-graphic-social-v3.gif.

37. Wilson, "The Story Behind 'Flatten the Curve.'"

38. "Members of the Coronavirus Task Force Hold a Press Briefing," YouTube video.

39. "Members of the Coronavirus Task Force Hold a Press Briefing."

40. Kyff, "Word Watch: Flatten the Curve." Nonhuman animals, in all their distinctiveness and variety, have long offered convenient visual references. See Hawhee, *Rhetoric in Tooth and Claw,* 121.

41. Billboard story: Swant, "Reddit Cofounder Alexis Ohanian Bought A Billboard."

Pledge: MoveOn.org, "Commit to #FlattenTheCurve." US Surgeon General, "A little #MondayMotivation."

42. Lai and Collins, "Which Country Has Flattened the Curve."

43. Breteau, "Coronavirus: Visualisez les pays." Language note: "aplatir la courbe" (infinitive form) is the common form of the slogan in France.

44. One particularly popular site, flattenthecurve.com, was created and maintained by Julie McMurray, a public health researcher in Oregon.

45. Bromfield and McConnell, "Two Routes to Precarious Success," 518–35. Coscieme et al., "Women in Power." Jamieson, "'Go Hard, Go Early,'" 598–605.

46. Harris, "Flattening the Coronavirus Curve."

47. Hariman and Lucaites, "Icons, Appropriation, and the Co-Production of Meaning," 286.

48. Finnegan, *Making Photography Matter*, 1.

49. Gries, *Still Life with Rhetoric*.

50. Tufte, *Visual Explanations*, 49.

51. Birgit Schneider and Thomas Nocke refer to the hockey stick graph as "one of climate change's iconic images." Schneider and Nocke, "Image Politics of Climate Change," 17.

52. See my discussion of the work on magnitude in this book's introduction. Farrell's meditations on magnitude are worth examining in full, but perhaps his most quotable (and most often quoted) line on how magnitude works rhetorically is this: "In short, magnitude says . . . 'Hey look at this! This is important!'" Farrell, "The Weight of Rhetoric," 484. Finnegan, too, quotes this line from Farrell in her development of the concept (Finnegan, *Making Photography Matter*, 130) as does Olson (in *American Magnitude*, 18).

53. Roberts, "Flattening the Coronavirus Curve."

54. Tufte, *The Visual Display of Quantitative Information*, 154.

55. Toby Morris, the illustrator in New Zealand, worrying that those colors in particular might compromise the message because of their association with political parties, changed them to teal and mustard yellow. Wilson, "The Story Behind 'Flatten the Curve.'" For an insightful discussion of color and political, sensory association, as well as a discussion of the red MAGA hats, see Poole, "Orientation: Seeing and Sensing Rhetorically," 14–15.

56. Hayes, "Flatten the (Climate) Curve." Hayes would later serve as Special Assistant to President Joe Biden for Climate Policy.

57. Shaw, "The Two Degrees Celsius Limit."

58. Hayes, "Flatten the (Climate) Curve."

59. Leiserowitz and Smith, "Affective Imagery," 3; and Leiserowitz and Smith, "The Rise of Global Warming Skepticism," 1021–32.

60. Gries, *Still Life with Rhetoric*, 62.

61. The technical summary of the 2021 IPCC report provides examples here. P. A. Arias et al., "Technical Summary," TS 45; TS 55; TS-121; TS-127; TS-130.

62. During the summer of 2020, my Google research alerts, set to notify me when the words "flatten," "climate," and "curve" showed up together in new places, pinged me multiple times weekly.

63. This issue was published days before the image of a raised Black fist resurged as a symbol in the United States and around the world after the police killing of George Floyd, an unarmed Black man, in Minnesota on May 25, 2020.

64. For more on the smokestack as one among many commonplace images ("iconic" is the author's term) helping people visualize the causes of climate change, see O'Neill, "Image Matters," 14, 16.

65. "Seize the Moment," *The Economist*, 7; "Flattening the Other Curve," *The Economist*, 14–15.

66. Howard Markel, a physician and historian of medicine at the University of Michi-

gan, first uttered the phrase "flatten the curve" over dinner with Marty Cetron, director of the CDC's Division of the Global Migration and Quarantine, back in 2005. Gounder, "A Second Wave?"

67. Sources that use the phrase "bend the curve" include: Mace et al., "Aiming Higher to Bend the Curve of Biodiversity Loss," 448–51; Raskin, "Bending the Curve: Toward Global Sustainability," 67–74. In public health scholarship on Ebola: Bailey et al., "Bending the Curve: Force Health Protection during the Insertion Phase of the Ebola Outbreak Response," 191. On cardiovascular disease: Fihn, "Bending the Curve on Cardiovascular Risk," 48–50. On tuberculosis: World Health Organization, Regional Office for South-East Asia, "Tuberculosis, 'Bending the Curve.'" Also Doshi, "Bending the Curve on Human Immunodeficiency Virus Transmission," 301–3.

68. António Guterres, "Statement by the Secretary-General."

69. Branswell, "Why 'Flattening the Curve' May Be the World's Best Bet."

70. Ramanathan et al., *Bending the Curve: Climate Change Solutions*, ix.

71. Ramanathan et al., viii.

72. Ramanathan et al., xiv.

73. Ramanathan et al., 4.

74. Ramanathan et al.: Keeling Curve, fig. 1.2.6, 1–19; Global Land-Ocean Temperature Curve, fig. 1.3.2, 1–29; Box 1.4.3, "Dangerous to Existential Risk: Categories of Projected 2100 Warming," 1–45; fig. 4.1.1, "Projections of Future Warming, Showing Business-as-Usual and Pathways for CO_2 Mitigation Only, SLCP [short-lived climate pollutant] Mitigation Only, and Mitigation of Both CO_2 and SLCP 4-10."

75. *Oxford English Dictionary*, s.v., "inflection," meaning 3 (Geometry).

76. *Oxford English Dictionary*, s.v., "inflection," meaning 1.c *figurative*.

77. *Oxford English Dictionary*, s.v., "bend, v.," meaning 7a.

78. *Oxford English Dictionary*, s.v., "flatten, v."

79. It is possible, too, that the stress on the word "flatten," without early and robust governmental restrictions, made the action into matter of individual effort.

80. As in: Fifield, "New Zealand Isn't Just Flattening the Curve"; Mandhana, "Some Countries Are Squashing the Coronavirus Curve"; BBC News, "Coronavirus: How South Korea Crushed the Curve."

81. This version is reprinted from MacCarthy et al., *Climate Change 2001*, 49.

The graph appears in the full report three times: in the Summary for Policy-Makers, in the technical summary, and then in the full report itself: Intergovernmental Panel on Climate Change, *Climate Change 2001*, 3, 29, 134.

82. Shapiro, "Michael Mann on Climate Change."

83. MacCarthy et al., *Change 2001*, 49.

84. Marcott et al., "A Reconstruction of Regional and Global Temperature," 1198.

85. Theo Stein, "Rise of Carbon Dioxide Unabated."

86. The interactive feature accompanying the 2019 UN Emissions Gap report notes that "temperatures have already increased 1.1°C." United Nations, "Emissions Gap Report 2019." Interactive feature: https://www.unenvironment.org/interactive/emissions-gap -report/2019/.

87. A small industry has grown up in rhetoric around Gore's documentary. See, for example, Lyons, "'Gore Is the World,'" 1156–70; Rosteck and Frentz. "Myth and Multiple Readings," 1–19; Johnson, "(Environmental) Rhetorics," 29–46; Olson, "Rhetorical Leadership and Transferable Lessons," 90–109; Aaltonen, "Claims of Hope and Disasters," 61–75.

88. Rosteck and Frentz mention such gasps, as does Mann, who tells Chris Mooney of

the *Atlantic* that he often hears "an audible gasp" when he shows the graph in his lectures. Mooney, "The Hockey Stick"; Rosteck and Frentz, "Myth and Multiple Readings," 2.

89. Olson, "American Magnitude," 380–404, and *American Magnitude*, 180–82; Rice, "The Rhetorical Aesthetics of More"; Larson, "'Just Let This Sink in,'" 432–44. As I mentioned at this chapter's outset, chapter 4 of Finnegan's *Making Photography Matter* treats magnitude at length, and my previous book, *Rhetoric in Tooth and Claw*, shows how nonhuman animals circulate magnitude.

90. Farrell, "The Weight of Rhetoric," 471.

91. *Oxford English Dictionary*, s.v., "order, n.," meaning 4b, "order of magnitude." Joshua DiCaglio discusses orders of magnitude in relation to scale—DiCaglio, *Scale Theory*, 61–62, 234.

92. Tufte, *The Visual Display of Quantitative Information*, 87.

93. Tufte, 87.

94. Tufte, *Visual Explanations*, 13.

95. Tufte, *The Visual Display of Quantitative Information*, 161.

96. Jenny Rice makes a similar point vis-à-vis big data: "in the case of Big Data, we may better understand data to be coherent not because of its contents but because it has literally made *sense*. It has given us a sense of the whole, yet the immensity of its size may also give us pause." Rice, "The Rhetorical Aesthetics of More," 46.

97. "Poetics," 7.1450b.34–1451a.4, in *Aristotle*, trans. Stephen Halliwell, 55, 57.

98. Peters, "33 + 1 Vignettes," 310.

99. Elsewhere, with different context and purpose, I write about the compensatory work of *phantasia*. Hawhee, *Rhetoric in Tooth and Claw*, 113–32.

100. See, for example, Kostelnick, "The Visual Rhetoric of Data Displays," 116–30; Kostelnick and Hassett, *Shaping Information*; Verhulsdonck and Shah, "Lean Data Visualization," 57–64; Van Winkle, "Above All Made by Themselves," 1–16. Work by Lydia Byrne, Daniel Angus, and Janet Wiles tries to empirically account for "socially constructed" meaning that viewers/readers bring to infographics—what they call "acquired codes of meaning"—but this approach, while incredibly useful for moving past semiotic approaches to data visualizations, does not encompass the convergence of mathematical magnitude with experiential, in-the-moment magnitude that I am getting at. See Byrne, Angus, and Wiles. "Acquired Codes of Meaning," 509–18.

101. Finnegan, *Making Photography Matter*, 158–62.

102. Larson, "'Just Let This Sink in,'" 433, 436.

103. Balzotti and Crosby, "Diocletian's Victory Column," 324. Also quoted in Larson, "'Just Let This Sink In,'" 433.

104. Balzotti and Crosby, "Diocletian's Victory Column," 330.

105. Lewis and Short, *A Latin Dictionary*, s.v., "amplification" and "amplifico," 110; Liddell, Scott, and Jones, *Greek-English Lexicon*, s.v., "ἐμφαίνω," 549. Jeanne Fahnestock mentions amplification in both of her books: Fahnestock, *Rhetorical Style*, 390–95; and Fahnestock, *Rhetorical Figures in Science*, 9, 12–13.

106. Millman, "Midtown Lawyer Positive for Coronavirus"; Rothfield et al., "13 Deaths in a Day."

107. An article published during the early part of the pandemic about exponential growth bias (EGB) specifically notes the repeated downplaying of numbers and display of EGB by US President Donald Trump and Brazilian President Jair Bolsonaro. Lammers, Crusius, and Gast, "Correcting Misperceptions of Exponential Coronavirus Growth," 16264.

108. Kunreuther and Slovic, "What the Coronavirus Curve Teaches Us."

109. Indeed, computer-generated models are designed to handle what mental models can't.

110. Houser, *Infowhelm*, 13.

111. Levy and Tasoff, "Exponential-Growth Bias," 1. As the pioneering environmental scientist Donella H. Meadows once observed, "The reason environmentalists are often so gloomy is that they know what the word 'exponential' means." Meadows, "Nothing Is so Powerful as an Exponential Whose Time Has Come."

112. One study compared the two and in fact found subjects' estimates more conservative when they encountered graphs instead of straight numbers. Wagenaar and Sagaria, "Misperception of Exponential Growth," 416–22.

113. Harris, "Flattening the Coronavirus Curve."

114. Gernot Wagner, "It's amazing to see how Covid-19 is climate change on warp speed." Quoted in "What Can We Learn from the Coronavirus Outbreak that Could Help Us Fight Climate Change?"

115. Intergovernmental Panel on Climate Change, IPCC 2018: Global Warming of 1.5°C, 190.

116. Hansen and Makiko, "Greenhouse Gas Growth Rates," 16109–14.

117. Hogg, Fonoberova, and Mezić, "Exponentially Decaying Modes," 16313.

118. "Seize the Moment," Editors, *The Economist*.

119. Torcello and Mann, "Seeing the COVID-19 Crisis."

120. Mann, *The New Climate War*, 238.

121. Ailsa Chang, "Interview with Brad Little."

122. Kolbert, "How Iceland Beat the Coronavirus."

123. Poggioli, "Lockdown, Safety Measures Bear Success in Italy."

124. Poggioli, "Lockdown, Safety Measures Bear Success in Italy."

125. Anderson, "Foreword," xiii.

126. Anderson, xiii.

127. McKibben, "130 Degrees," emphasis added.

128. Olson, *American Magnitude*, 23–25. 180–88. Rice's *Awful Archives* is also a rich resource on how conspiracy depends, circulates, and is amplified by magnitude. The question of multiple magnitudes circulating and, in some cases, competing deserves further investigation.

129. Again, Olson admirably documents and reflects on Donald Trump's particular version of American magnitude and its accompanying dangers—Olson, *American Magnitude*, 180–90. The only time Trump's version of American magnitude aided the pandemic was in the context of Operation Warp Speed, the administration's name for the quest to develop a vaccine.

130. The idea of irreconcilable magnitudes leading to, or even necessitating, denial is one of the implications of Stephanie Larson's work, so I owe her thanks here. Indeed, though it does not say so outright, her work has nevertheless convinced me that the denialist logics of rape culture may well pervade *all* denialist discourse. See, in particular, the introduction to her first book, Larson, *What It Feels Like*.

Chapter Five

1. Madison Square Park Conservancy, "Step into Our Tree Museum."

2. The artist Maya Lin tells Evan Moffitt in a *Frieze* interview that the trees were one or two years away from certain death, but that for safety reasons she had to use trees that were still clinging to life. Moffitt, "Maya Lin Builds a Final Monument to the Climate."

3. Lin explained these matters at a small press preview on opening day, which I attended at curator Brooke Kamin Rapaport's generous invitation. Lin, "Press Preview Remarks." The art journalist Adrian Horton's *Guardian* piece on the installation expands on Lin's remarks. See Horton, "'Odd, Eerie and Haunting.'"

4. Rapaport, "Maya Lin's 'Ghost Forest.'"

5. Pezzullo and Cox, *Environmental Communication and the Public Sphere*, 72.

6. I want to be clear that protest signs (and the demonstrations of which they are a part) and if-then arguments of course remain useful in particular contexts. The point in this book is to identify alternative forms of conveying urgency.

7. The work in rhetorical studies on publics is vast. For influential works that themselves provide useful bibliographies, see Asen, "Public: A Network of Relationships," 297–305; Rice, *Distant Publics*; Ackerman and Coogan, *The Public Work of Rhetoric*; and Goodnight, "The Personal, Technical, and Public Spheres of Argument," 214–27.

8. For foundational work on public memorials, see Blair, "Contemporary US Memorial Sites," 16–57; Doss, *Memorial Mania*; Blair, Balthrop, and Michel, "Mood of the Material," 6–20; and Blair, Jeppeson, and Pucci, "Public Memorializing in Postmodernity," 263–88. For an illuminating consideration of the National September 11 memorial, see Vivian, *Commonplace Witnessing*, 129–58; and Phillips, "The Failure of Memory," 208–23. On the distinction between monuments and memorials, see Sturken, "The Wall, the Screen, and the Image," 120.

9. Margaret LaWare and Victoria Gallagher's framework for approaching rhetoric and public art, for example, values the "promise of public, democratic deliberation" held by public art. LaWare and Gallagher, "The Power of Agency," 162. Candice Rai considers how "[p]ublic art and its contradictory purposes bring to the fore a series of productive tensions that exist in the liberal democratic public, more broadly"—Rai, *Democracy's Lot*, 115. And Caitlin Bruce incorporates those democratic, civic commitments into her very definition of public art as "a creative practice that either generates or is framed by discourse about shared space and collective life"—Bruce, *Painting Publics*, 15.

10. Jason Peters focuses on the educational engagement fostered by community-produced art, showing how artistic method and making itself can deepen awareness of local environmental issues. Peters, "Public Art as Social Infrastructure," 106–29. The rhetoric scholar Elizabeth Brunner does not write about public art per se, but her observations about environmental art seem to me applicable here, particularly this one: "Art is used to spread information, document abuses and grandeur, and provoke contemplation by encouraging new ways of considering degradation, crises, and protection"—Brunner, "Contemporary Environmental Art in China," 405.

11. Song, "A New Vision for Public Art," 244.

12. Douglas, "Moderator Remarks."

13. Thanks once again to Casey Boyle, who read an earlier draft of this chapter and encouraged me to draw out—perhaps paradoxically—this obliqueness.

14. DeLaure, "Performing Loss"; Kolodziejski, "What Is Missing?," 428–45.

15. "Ghost Forest—Madison Square Park," *What Is Missing*? These headings do not simply duplicate the information presented in the "Ghost Forest" soundscape. In fact, the *What Is Missing* site spotlights a few success stories, such as the return of the harbor seal. And a search for the phrase "Ghost Forest" yields a clutch of subheadings (sixteen as of June 21, 2021), each one spotlighting ghost forests in different locations around the world. https://whatismissing.org/timeline/ghost-forest-climate-change.

16. Blair's "Contemporary US Memorial Sites" presents the Vietnam Veterans Memorial as one of five exemplars. See also Blair, Jeppeson, and Pucci, "Public Memorializing."

17. Blair and Neil Michel also write about Lin's Civil Rights Memorial in Montgomery, Alabama, in "Reproducing Civil Rights Tactics," 31–55.

18. Blair, "Contemporary US Memorial Sites," 17.

19. Grosz, Chaos, Territory, Art, 4.

20. Blair discusses the "recalcitrant presentness" of the memorials she examines in "Contemporary US Memorial Sites," 17, though thanks in many ways to her foundational work, I am at this point more curious about the enlivening of senses through that presentness.

21. Hawhee, "Rhetoric's Sensorium," 12–13.

22. Wysocki, "Unfitting Beauties of Transducing Bodies," 94. Special thanks to Casey Boyle for suggesting the fruitful connection to Wysocki's work.

23. This phrase appears at least fourteen times in Perelman and Olbrechts-Tyteca's book and twice in Perelman's single-authored book: Perelman and Olbrechts-Tyteca, The New Rhetoric, 118, 158, 160, 161, 163, 175 (twice), 176, 504. Their discussion of tense even lays claim to the phrase: "The present has the further property of conveying most readily what we have called 'the feeling of presence'" (160). See also Perelman, The New Rhetoric and the Humanities, 18, 19.

24. Perelman and Olbrechts-Tyteca, The New Rhetoric, 117. They later add that "Presence, and efforts to increase the feeling of presence, must hence not be confused with fidelity to reality" (118).

25. Pezzullo, Toxic Tourism, 9–10; Kennerly, "Getting Carried Away," 269. Cara Finnegan, working in the context of the photographic form, points out the sensuous aspect of presence, from which she builds an account of photography as a form of "virtual witnessing." See Finnegan, Making Photography Matter, 14–15, 27.

26. Hawhee and Poole, "Kenneth Burke at the MoMA," 428.

27. Once I knew what to look for, I was better able to make out the installation from most points while walking around the outskirts of the park.

28. "Maya Lin Ghost Forest," Placard, visited on May 10, 2021.

29. Helfand, "Ghost Forest," 18.

30. Schroeder, Taras, and United States Forest Service, Atlantic White-Cedar, 4.

31. Apart from that which is already cited, information in this paragraph comes from two sources: US Department of Agriculture and Natural Resources Conservation Service, "New Jersey Fact Sheet: Atlantic White Cedar"; and Byers, "Restoring the 'Kidneys of the Pine Barrens.'" Philip Deloria writes that the "world took as its material base the accumulation of capital ripped from indigenous lands, resources, and labor over the course of the centuries"—Deloria, Indians in Unexpected Places, 23.

32. Tirmenstein, "Chamaecyparis Thyoides in Fire Effects Information System."

33. McPhee, The Pine Barrens, 105.

34. According to a Forest Service report, the trees reach greater heights in the more southern locations. So, for example, in South Carolina, they can grow to 120 feet and reach a diameter of 5 feet. But in New Jersey, mature trees range between 40 and 60 feet. Schroeder, Taras, and US Forest Service, Atlantic White-Cedar, 2.

35. Tirmenstein, "Chamaecyparis Thyoides in Fire Effects Information System."

36. Schroeder, Taras, and US Forest Service, Atlantic White-Cedar, 2.

37. McPhee, The Pine Barrens, 105. The marine scientists Matthew L. Kirwan and Keryn B. Gedan, in a review of research on ghost forests, note that "in Delaware Bay, Atlantic white cedar (Chamaecyparis thyoides) forests died back at faster rates than hardwood forests." Kirwan and Gedan, "Sea-Level Driven Land Conversion," 452.

38. Kummer, "Rising Seas." It should be said that ghost forests are a global phenome-

non, and they include many kinds of tree species—Ponderosa pines in the western US, for example, are vulnerable to beetle infestations when winter temperatures do not remain low enough.

39. Parry, "Seas Rise, Trees Die."

40. Drouin, "How Rising Seas Are Killing Southern US Woodlands."

41. Root, "Ghost Forests Are Visceral Examples."

42. Schieder and Kirwan, "Sea-Level Driven Acceleration," 1154.

43. Root, "Ghost Forests Are Visceral Examples."

44. National Oceanographic and Atmospheric Administration, "What Is a Ghost Forest?"

45. Two scholars in rhetoric have been working on concepts related to what I am here calling felt science. Megan Poole studies the punctive force of scientific inquiry—how aesthetics, with an emphasis on sensation and feeling, figures into, even at times drives, that age-old empirical category of "discovery." Poole, "Technical Beauty." And Lynda Walsh's book *Scientists as Prophets*, which I considered in chapter 4, documents the twentieth-century elevation of scientists to the status of prophets, their expertise allowing them to see and attest to futures more clearly than those not equipped with that expertise. Walsh's chapter on climate change and prophetic technologies of modeling documents the disciplining of uncertainty, but it also contains an undercurrent of feeling. Walsh, *Scientists as Prophets*, 163–85. I owe a debt to both of these rhetoric and science scholars for providing such clear examples of what I am discussing here.

46. Kummer, "Rising Seas."

47. Root, "Ghost Forests Are Visceral Examples."

48. Pogrebin, "Maya Lin's 'Ghost Forest' Is a Climate Warning," 3.

49. Pogrebin.

50. Madison Square Park Conservancy, "Press Release: Maya Lin Ghost Forest."

51. Bury, "Gray and Green."

52. Small, "Maya Lin's Dismantled 'Ghost Forest.'"

53. Pogrebin, "Maya Lin's 'Ghost Forest' Is a Climate Warning."

54. Rapaport, "Maya Lin's 'Ghost Forest.'"

55. Aoki, Dickinson, and Ott, "The Master Naturalist Imagined," 246–49. See also Hawhee and Poole, "Kenneth Burke at the MoMA," 421; Wysocki, "Unfitting Beauty," see above for discussion.

56. Balzotti and Crosby, "Diocletian's Victory Column," 330. See chapter 4 for a more thorough discussion of this article.

57. Helfand, "Ghost Forest," 18.

58. Madison Square Park Conservancy, "Ghost Forest Full Documentary."

59. Bury, "Gray and Green."

60. As the dendrochronologist Valerie Trouet explains, spiral growth in conifers can result from "a variety of stress factors (such as asymmetrical crowns, wind, and slope)"— Trouet, *Tree Story*, 30. See also Schulgasser and Witztum, "The Mechanism of Spiral Grain Formation in Trees," 133–56.

61. Maya Lin, "Press Preview Remarks."

62. Madison Square Park Conservancy, "Garden Jewels, April 12–30."

63. Rapaport, "Maya Lin's 'Ghost Forest," emphasis added to reflect the audible stress Rapaport places on the word "next."

64. Helfand, "Ghost Forest."

65. Goodman, "$2.3B Office Tower Breaks Ground."

66. Gordon, "Some Thoughts on Haunting and Futurity," 2.

67. Key among work on haunting in rhetoric is that by Joshua Gunn. See especially Gunn, "Mourning Speech," 91–114. Gunn's review essay on haunting did the field the important service of drawing in Gordon's work, which figures prominently in my analysis. Gunn, "Review Essay: Mourning Humanism," 77–102. Derrida's work on hauntology dominates rhetoric, as it does most work on haunting in the humanities. Gordon's sociological work and Tuck and Ree's postcolonial perspective offer a more apt and generative point of connection than does Derrida for Lin's art. Gordon, *Ghostly Matters*; and Tuck and Ree, "A Glossary of Haunting," 639–58. But for a relevant and profound treatment of extinction's spectrality in the Derridean vein, see Whale and Ginn, "In the Absence of Sparrows," 92–116.

68. Murphy, "Presence, Analogy, and Earth in the Balance," 5.

69. Tuck and Ree, "A Glossary of Haunting," 642.

70. LaBelle, *Acoustic Justice*, 86–87.

71. LaBelle, 85.

72. The land acknowlegment: "Madison Square Park is located on Lenapehoking, the ancestral homeland of The Lenape (Delaware) people. We recognize that this land was forcibly taken, resulting in the displacement and genocide of the Lenape (Delaware) Nations. Madison Square Park Conservancy respectfully acknowledges the Lenape (Delaware) peoples—past, present, and future—who continue to live, work, and connect to this land." Madison Square Park Conservancy, "Maya Lin *Ghost Forest* Soundscape." See also Madison Square Park Conservancy, "Maya Lin Conceives a Soundscape for *Ghost Forest*."

73. Madison Square Park Conservancy, "Maya Lin Conceives a Soundscape for *Ghost Forest*."

74. Given that it was available on the Conservancy's website while the installation was open, and that it remains accessible on the *What Is Missing* website, the soundscape's conjuring work can stand alone. See "Ghost Forest," *What Is Missing?*

75. Madison Square Park Conservancy, "Maya Lin Conceives a Soundscape for *Ghost Forest*."

76. Leveille, "Artist Maya Lin's Sound Ring." The quotation in PRI's written extract omits the word "to" (so: "sound is our way in these species and places"), but the word "to" is clearly audible in the recording of Lin's remarks, and her pause between the word "in" and "to" led me to treat them as separate words rather than one ("into," as DeLaure for one renders it when she quotes this remark—DeLaure, "Performing Loss," 17). The shift in meaning from "way in to" to "way into" is subtle but important, in keeping with Lin's artistic tendencies.

77. Borgmann, "Sounds from the Macaulay Library."

78. Lewis and Short, *A Latin Dictionary*, s.v., "intimaro" and "intimus," 988.

79. Leveille, "Artist Maya Lin's Sound Ring."

80. Boyle, *Rhetoric as a Posthuman Practice*, 81–82. Byron Hawk, too, observes of sound and feeling: "As an embodied event, sound modulates feelings and behaviors in ways that aren't always immediately clear or tied to particular meanings." Hawk, *Resounding the Rhetorical*, 2. Boyle, Brown, and Ceraso present a lucid and compelling overview of rhetoric as transduction in "The Digital," 251–59.

81. Gallagher, "Field Recording and the Sounding of Spaces," 569.

82. Gallagher, 566, 572.

83. The whip-poor-will call can no longer be heard in those woods. Whip-poor-will populations have declined by almost 70 percent throughout their range. The Cornell Lab, "Are Whip-Poor-Will Populations Declining?"

84. Dyson, *Sounding New Media*, 4.

85. DeLaure, "Performing Loss," 13.

86. DeLaure, 13.

87. Aoki, Dickinson, and Ott, "The Master Naturalist," 258.

88. Gallagher, "Field Recording and the Sounding of Spaces," 569.

89. DeLaure discusses the haunting quality of nonhuman animal sounds in Lin's final memorial, focuses on nonhuman voices in relation to work on recordings of human voices, and identifies ways that haunting links to mourning—DeLaure, "Performing Loss," 20–24.

90. For a poignant account of rhetorical relations between naming and ecological mourning, see Barnett, "Naming, Mourning, and the Work of Earthly Coexistence," 287–99.

91. Vineyard Gazette, "Rare Footage of Heath Hen on Martha's Vineyard." See also "The Last Heath Hen," 382–84. A search of "Heath Hen" on Lin's *What Is Missing* site brings up excerpts from several primary sources documenting the bird's decline.

92. Madison Square Park Conservancy, "Maya Lin *Ghost Forest* Soundscape."

93. The literature on Traditional Ecological Knowledge (TEK) is vast. The indigenous philosopher Kyle Powys Whyte offers a helpful overview of the various and sometimes conflicting ways the concept has been used and makes a case for focusing on coordination. Whyte, "On the Role of Traditional Ecological Knowledge," 7. In his introduction to the edited volume on TEK, Dan Shilling frames the matter in this way: "Ironically, it has taken decades for many sustainability scholars to consider the spiritual and ecological underpinnings that helped define native peoples' relationship to nature, even though these relationships 'sustained' most tribal nations, and the land on which they lived, for thousands of years." See Shilling, "Introduction: The Soul of Sustainability," 3–4.

94. Tuck and Ree, "A Glossary of Haunting," 643.

95. Barnett's "Naming, Mourning, and the Work of Earthly Coexistence" helps bolster the case for haunting as a constituent element of the climate crisis.

96. Radway, "Foreword," x.

97. Thomas Rickert has most fruitfully elaborated the idea of attunement for rhetoric, and his "ambient rhetoric" inhabits sonic modalities with theoretical precision. Rickert, *Ambient Rhetoric*, 8, 130–57. Byron Hawk (*Resounding the Rhetorical*, 187–223) foregrounds sound and attunement as well. See in particular his explanation of attunement as a kind of ambient tuning (189). He arrives at "a resonant sense of the rhetorical" that I find especially generative. Trevor Hoag, too, presents the ghost as an aid to thinking and recalling, and he finds prudence in listening. See Hoag, "Ghosts of Memory," 3.

98. Small, "Maya Lin's Dismantled 'Ghost Forest.'"

99. Dimock, "Hearing Animals," 401.

Chapter Six

1. Sussman, *The Oldest Living Things in the World*.

2. In an essay included in the book, the curator and art historian Hans Ulrich Obrist writes, "Sussman does not only utilize scientific strategies and tools for the aesthetics of her work; she actually builds up an archive, consisting of the images from the organisms, diagrams, and maps. Therefore the product of the research develops into an archive that will be the foundation of what will travel into the future." Obrist, "The Future Is Invented," xvii.

3. Sussman, *The Oldest Living Things in the World*, xxiii.

4. Sussman, xxiii.

5. "Photography," writes Sussman in the book's preface, "is the ideal medium to capture the temporal tension: thousands of years of a life, distilled down to small fractions of a second." Sussman, x.

6. Sussman, "The Oldest Living Things in the World."

7. Sussman, *The Oldest Living Things in the World*, x.

8. Sussman. Though to be honest, one might come for the trees and stay for the map lichens of Greenland (pp. 66–71), the Antarctic moss (pp. 240–49), and the astonishing Siberian Actinobacteria continuously living in permafrost for between 400,000 and 600,000 years (pp. 168–171).

9. Sussman: Fortingall Yew, 112; Antarctic beech, 201–7.

10. Sussman, 3–7. The dendrochronologist Nate Stephenson told Sussman "that in fact *hundreds* of the trees could be over 2,000, but that they don't have the resources to date them all" (6).

11. *Oxford English Dictionary Online*, s.v., "fathom, v."

12. "The fathom, one of the most evocative of nautical words, is to disappear from British Admiralty Charts," reads the first line of a Manchester news report, "Full Metre Nine Thy Father Lies," 1. This article is listed in the entry for the noun form of the word. *Oxford English Dictionary Online*, s.v., "fathom, n."

13. The idea of the Anthropocene has received much cross-disciplinary scrutiny. Standard citations include Steffen, Crutzen, and McNeill, "The Anthropocene," 614–21; and Lewis and Maslin, "Defining the Anthropocene," 171–80. In addition to Keller's work mentioned in the Introduction, I appreciate, in particular, Jennifer Clary-Lemon's consideration of the name in the introduction to *Planting the Anthropocene*, 5–6. I also wish to call attention to the question "whose Anthropocene?" posed by Eleanor Hayman, Colleen James, and Mark Wedge, in their "Future Rivers of the Anthropocene," 77–92. For a compelling elemental approach to the question of the Anthropocene, see Menely, "Anthropocene Air," 93–101.

14. Scott, *The Journal of Sir Walter Scott*, 538. Also cited in the *Oxford English Dictionary Online*, s.v., "fathom, v."

15. Martyr, *The Decades of the Newe Worlde*, II.68. The word "fathom" obviously carries residue of settler colonialism, and its related stripping of trees. So it is perhaps not surprising that the word was also used to quantify wood for building and burning. See *Oxford English Dictionary*, s.v., "fathom, n."

16. Wright, *Casting Deep Shade*, 37.

17. Am I suggesting that rhetoric works through such aids to the imagination? Yes I am. And not for the first time. For discussions of other rhetorical enhancements to the imagination, see Hawhee, *Rhetoric in Tooth and Claw*, especially chapters 3 and 5.

18. This epilogue is likely to be my first, not last, words on trees.

19. As Jason Edward Black observes, "trees are noticeable, if not abundant, in human world-making." Black, "Here Is a Strange and Bitter Crop," 27. Given that rhetoric intensifies most noticeably in a crisis, scholars have focused on moments when trees prompt controversy. And Jenny Rice, writing in the context of development rhetoric, acknowledges "the much larger social meanings that trees have had for millennia"—Rice, *Distant Publics*, 11. Rice's study of development rhetoric opens with the divisive removal of oak trees from the campus at the University of Texas at Austin to make room for a major renovation to the football stadium and includes a discussion of the poisoning of Austin's 500-year-old Treaty Oak (pp. 8–13). Madison Jones begins his study of how trees pull rhetoric beyond the human with the Auburn community's outpouring of feeling in response to the destruction of Toomer's Oaks on Auburn's campus. Jones calls such moments "sylvan rhetorics," which he defines as "the study of trees as relations for, and in relation to, rhetoric." Jones, "Sylvan Rhetorics," 63–65. Such rhetorics are as capacious as they are strange. Trees occupy an important place in memory scholarship as well, as Black's critical consideration of

the sycamore known as the Emmett Till Memorial Tree shows. David Maxson's brilliant dissertation on the lively memoryscape of New Orleans documents the concern that rose with flood waters through the Avenue of Oaks on the campus of Dillard University. The oaks themselves, according to Maxson, "act as revered witnesses that ground traditions in a perceptually stable landscape even as they mark the creeping passage of time through their growth." He calls them "active participants in living traditions." Maxson, "Rhetorical Burials," 119. I would also like to also acknowledge the rhetoric scholar Natalie Bennie's remarkable work on witness trees—and on one Tennessee witness tree in particular—in our 2020 graduate seminar on Rhetoric and Imagination. Trees hold great potency in relation to climate rhetoric in particular. Kevin DeLuca's important early work on environmental "image events" presents trees as an important part of such events and considers how they rearrange "human-nature relations"—DeLuca, *Image Politics*, 67–69, 8–10, 55–57. See also DeLuca and Demo, "Imagining Nature and Erasing Class and Race," 541–60. Jennifer Clary-Lemon's study of tree-planter discourse offers trees as "complex sticky bodies" that rouse mixtures of feelings. These feelings, Clary-Lemon finds, form intimate connections among human and nonhuman beings, tools, land, and energies (Clary-Lemon, *Planting the Anthropocene*, 113, 129). Joshua Trey Barnett's examination of an essay collection in which ecologists consider the demise of the Eastern Hemlock finds therein a mix of sentiment—what Barnett there and elsewhere terms "ecological grief"—and science. This point of Barnett's beautifully confirms what this book has shown about information and feeling in the case of one dying tree species. Barnett, "From Ecological Grief to Gelassenheit," 783–97.

20. Black, "Here Is a Strange and Bitter Crop," 28.

21. Barnett, "From Ecological Grief to Gelassenheit," 793. Clary-Lemon offers a compelling description of how planters relate to tree timescales, in *Planting the Anthropocene*, 92–97.

22. Trouet, *Tree Story*, 41.

23. Trouet puts it humorously in her book: "Like humans, old trees (more than 250 years old) look different from middle-aged trees (50–250 years), which look different from young trees (less than 50 years). Also like humans, trees grow in height only when they are young but continue to grow in girth as they age"—Trouet, *Tree Story*, 29.

24. Trouet, 6. I also appreciate Trouet's assertion "trees don't lie" (49), which perhaps implies another distinction from humans.

25. Trouet, 76–81.

26. Thunberg and Monbiot, "Greta Thunberg and George Monbiot Make Short Film on the Climate Crisis."

27. Government of Canada, "2 Billion Tree Commitment." Clary-Lemon's book examines industrial tree planting as a way that forests manage humans at least as much as the reverse. Clary-Lemon, *Planting the Anthropocene*, 5.

28. Horowitz, "Athens Is Only Getting Hotter"; Farhadi, Faizi, and Sanaieian, "Mitigating the Urban Heat Island," 101448; Middel, Chhetri, and Quay, "Urban Forestry and Cool Roofs," 178–86; Pozo et al., "Equity in Allocating Carbon Dioxide Removal Quotas," 640–46.

29. Maya Lin Studio, *Ghost Forest*, https://www.mayalinstudio.com/art/ghost-forest.

30. As just one example, the plant physiologist Angelica Patterson studies tree and plant response to increasingly warming air temperatures, predicting tree migration and adaptation—Patterson, "Warming Climates, Changing Forests," 160–71. And see Taylor et al., "Growth and Physiology of a Dominant Understory Shrub," 193–202. Patterson appeared as a guest on a podcast with Matt Candeias, "Trees in a Changing Climate."

31. Simard, "How Trees Talk to Each Other"; and Simard, *Finding the Mother Tree*.

See also Wohlleben, *The Heartbeat of Trees*. Simard's collaborators and Peter Wohlleben (a forester) have subsequently popularized Simard's work as the discovery of "the wood-wide web." See Helgason et al., "Ploughing Up the Wood-Wide Web?," 431; Wohlleben, *The Hidden Life of Trees*, 11.

32. Powers, *The Overstory*.

33. Grignon and Kimmerer, "Listening to the Forest," 73–74. For an example of a climate modeling study that integrates traditional ecological knowledge, particularly Diné kinship perspectives, see Yazzie et al., "Diné Kinship as a Framework," 1331–42. Also sharing stories of Menominee relations, Kyle Whyte, Chris Caldwell, and Marie Schaefer present qualities of relations that motivate care—Whyte, Caldwell, and Schaefer, "Indigenous Lessons about Sustainability," 160.

34. Strutt, *Sylva Britannica*, xv.

35. Strutt, 56.

36. Bernard-Donals and Jensen, "Introduction: Taking Rhetoric to Its Limits," 1, 3.

37. Kimmerer, *Braiding Sweetgrass*, 57.

38. Again, Bernard-Donals and Jensen note this bidirectional feature of the sacred at the outset of their discussion—Bernard-Donals and Jensen, "Introduction: Taking Rhetoric to Its Limits," 1.

39. See, for example, Bhagwat and Rutte, "Sacred Groves," 519–24; Khan, Khumbongmayum, and Tripathi, "The Sacred Groves and Their Significance," 277–91.

BIBLIOGRAPHY

Aaltonen, Jouko. "Claims of Hope and Disasters: Rhetoric Expression in Three Climate Change Documentaries." *Studies in Documentary Film* 8, no. 1 (January 2, 2014): 61–75.

Aarts, Bas. "Subjunctive." In *The Oxford Dictionary of English Grammar*. Oxford: Oxford University Press, 2014. https://www.oxfordreference.com/view/10.1093/acref/9780199658237.001.0001/acref-9780199658237-e-1437.

Ackerman, John, and David J. Coogan. *The Public Work of Rhetoric: Citizen-Scholars and Civil Engagement*. Studies in Rhetoric/Communication. Columbia: University of South Carolina Press, 2013.

Agamben, Giorgio. *Remnants of Auschwitz: The Witness and the Archive*. New York: Zone Books, 2002.

Alaimo, Stacy. *Bodily Natures: Science, Environment, and the Material Self*. Bloomington: Indiana University Press, 2010.

Alaimo, Stacy. *Exposed: Environmental Politics and Pleasures in Posthuman Times*. Minneapolis: University of Minnesota Press, 2016.

Allen, Thomas R., and John A. Kupfer. "Application of Spherical Statistics to Change Vector Analysis of Landsat Data: Southern Appalachian Spruce-Fir Forests." *Remote Sensing of Environment* 74, no. 3 (December 1, 2000): 482–93.

Allen, Thomas R., and John A. Kupfer. "Spectral Response and Spatial Pattern of Fraser Fir Mortality and Regeneration, Great Smoky Mountains, USA." *Plant Ecology* 156, no. 1 (September 1, 2001): 59–74.

Alter, Charlotte, Suyin Haynes, and Justin Worland. "Person of the Year: The Conscience." *Time*, December 23 and 30, 2019, 55.

Aristotle. *On Rhetoric: A Theory of Civic Discourse*. Translated by George A. Kennedy. 2nd edition. New York: Oxford University Press, 2007.

Anderson, Inger. "Foreword." In *Emissions Gap Report 2019*, 13. Nairobi: United Nations Environmental Programme, 2019.

Aoki, Eric, Greg Dickinson, and Brian L. Ott. "The Master Naturalist Imagined: Directed Movement and Simulations at the Draper Museum of Natural History." In *Places of Public Memory: The Rhetoric of Museums and Memorials*, ed. Greg Dickinson, Carole Blair, and Brian L. Ott, 238–65. Tuscaloosa: University of Alabama Press, 2010.

Appiah, Kwame Anthony. *As If: Idealization and Ideals.* Cambridge, MA: Harvard University Press, 2017.

Ardern, Jacinda. "Prime Minister Jacinda Ardern Speaks After Coronavirus Committee Meeting." Newshub, March 14, 2020. https://www.newshub.co.nz/home/politics /2020/03/livestream-prime-minister-jacinda-ardern-speaks-after-coronavirus -committee-meeting.html.

Arias, P. A., N. Bellouin, E. Coppola, R. G. Jones, G. Krinner, J. Marotzke, V. Naik, M. D. Palmer, G.-K. Plattner, J. Rogelj, M. Rojas, J. Sillmann, T. Storelvmo, P. W. Thorne, B. Trewin, K. Achuta Rao, B. Adhikary, R. P. Allan, K. Armour, G. Bala, R. Barimalala, S. Berger, J. G. Canadell, C. Cassou, A. Cherchi, W. Collins, W. D. Collins, S. L. Connors, S. Corti, F. Cruz, F. J. Dentener, C. Dereczynski, A. Di Luca, A. Diongue Niang, F. J. Doblas-Reyes, A. Dosio, H. Douville, F. Engelbrecht, V. Ering, E. Fischer, P. Forster, B. Fox-Kemper, J. S. Fuglestvedt, J. C. Fyfe, N. P. Gillett, L. Goldfarb, I. Gorodetskaya, J. M. Gutierrez, R. Hamdi, E. Hawkins, H. T. Hewitt, P. Hope, A. S. Islam, C. Jones, D. S. Kaufman, R. E. Kopp, Y. Kosaka, J. Kossin, S. Krakovska, J.-Y. Lee, J. Li, T. Mauritsen, T. K. Maycock, M. Meinshausen, S.-K. Min, P. M. S. Monteiro, T. Ngo-Duc, F. Otto, I. Pinto, A. Pirani, K. Raghavan, R. Ranasinghe, A. C. Ruane, L. Ruiz, J.-B. Sallée, B. H. Samset, S. Sathyendranath, S. I. Seneviratne, A. A. Sörensson, S. Szopa, I. Takayabu, A.-M. Tréguier, B. van den Hurk, R. Vautard, K. von Schuckmann, S. Zaehle, X. Zhang, and K. Zickfeld. "Technical Summary." In *Climate Change 2021: The Physical Science Basis.* Contribution of Working Group I to the Sixth Assessment Report of the Intergovernmental Panel on Climate Change, edited by V. Masson-Delmotte, P. Zhai, A. Pirani, S. L. Connors, C. Péan, S. Berger, N. Caud, Y. Chen, L. Goldfarb, M. I. Gomis, M. Huang, K. Leitzell, E. Lonnoy, J. B. R. Matthews, T. K. Maycock, T. Waterfield, O. Yelekçi, R. Yu, and B. Zhou. Cambridge: Cambridge University Press, 2021.

Aristotle. "Poetics." In *Aristotle*, trans. Stephen Halliwell, vol. 23, 28–141. Loeb Classical Library 199. Cambridge, MA: Harvard University Press, 1995.

Asen, Robert. "Public: A Network of Relationships." *Rhetoric Society Quarterly* 48, no. 3 (May 27, 2018): 297–305.

Asmelash, Lee. "Greta Thunberg Has a Suggestion for Congress on How to Take Real Action on the Climate Crisis." CNN Politics, September 18, 2019. https://www.cnn .com/2019/09/18/politics/greta-thunberg-climate-hearing-congress-trnd/index .html.

Australian Academy of Science. "The Australian Bushfires: Why They Are Unprecedented." Australian Academy of Science, February 3, 2020. https://www.science .org.au/news-and-events/news-and-media-releases/australian-bushfires-why-they -are-unprecedented.

Bailey, Mark S., K. Beaton, D. Bowley, W. Eardley, P. Hunt, S. Johnson, J. Round, N. T. Tarmey, and A. Williams. "Bending the Curve: Force Health Protection during the Insertion Phase of the Ebola Outbreak Response." *Journal of the Royal Army Medical Corps* 162, no. 3 (June 1, 2016): 191–97.

Bakhtin, Mikhail. *Dialogic Imagination: Four Essays.* Austin: University of Texas Press, 1981.

Ballif, Michelle. "Divining Rhetoric's Future." In *Responding to the Sacred: An Inquiry Into the Limits of Rhetoric*, edited by Michael Bernard-Donals and Kyle Jensen, 157–71. University Park: Pennsylvania State University Press, 2021.

Balthrop, V. William, Carole Blair, and Neil Michel. "The Presence of the Present:

Hijacking 'The Good War'?" *Western Journal of Communication* 74, no. 2 (April 1, 2010): 170–207.

Balzotti, Jonathan Mark, and Richard Benjamin Crosby. "Diocletian's Victory Column: Megethos and the Rhetoric of Spectacular Disruption." *Rhetoric Society Quarterly* 44, no. 4 (August 8, 2014): 323–42.

Bangerter, Adrian, and Max M. Lourwerse. "Focusing Attention with Deictic Gestures and Linguistic Expressions." *Proceedings of the Annual Meeting of the Cognitive Science Society* 27 (2005): 1331–36.

Banning, Marlia E. "Exigency in Dispute: Global Warming and Uncertainty in Contemporary Rhetorical Criticism." *JAC* 31, no. 3/4 (2011): 638–60.

Barnett, Joshua Trey. "From Ecological Grief to Gelassenheit: Rhetorics of Sentiment and Science in Hemlock." *Environmental Communication* 15, no. 6 (April 6, 2021): 783–97.

Barnett, Joshua Trey. "Grievable Water: Mourning the Animas River." In *Mighty Stream: A Confluence of Water, Rhetoric, and Social Justice*, 273–91. Lanham, MD: Lexington Press, 2020.

Barnett, Joshua Trey. *Mourning in the Anthropocene: Ecological Grief and Earthly Coexistence*. East Lansing: Michigan State University Press, 2022.

Barnett, Joshua Trey. "Naming, Mourning, and the Work of Earthly Coexistence." *Environmental Communication* 13, no. 3 (April 3, 2019): 287–99.

Barnett, Joshua Trey. "Vigilant Mourning and the Future of Earthly Coexistence." In *Communicating in the Anthropocene*, 13–33. Lanham, MD: Lexington Books, 2021.

Barrett, Vic. "Yes, I'm Striking Over the Climate Crisis, and Suing the US Government, Too." *Guardian*, September 20, 2019. https://www.theguardian.com/commentisfree/2019/sep/20/climate-crisis-why-i-am-suing-the-us-government.

Bastian, Michelle. "Fatally Confused: Telling the Time in the Midst of Ecological Crises." *Environmental Philosophy* 9, no. 1 (2012): 23–48.

Bazerman, Charles. "Afterword: Social Changes in Science Communication: Rattling the Information Chain." In *Science and the Internet: Communicating Knowledge in a Digital Age*, edited by Alan G. Gross and Jonathan Buehl, 267–82. Amityville, NY: Baywood Publishing Company, Inc., 2016.

BBC News. "Coronavirus: How South Korea Crushed the Curve." May 10, 2020. https://www.bbc.com/news/av/world-asia-52584494.

Bellacasa, Maria Puig de la. *Matters of Care: Speculative Ethics in More than Human Worlds*. PostHumanities (Paperback) Series. Minneapolis: University of Minnesota Press, 2017.

Berlant, Lauren. "Trauma and Ineloquence." *Cultural Values* 5, no. 1 (January 1, 2001): 41–58.

Bernard-Donals, Michael. *Forgetful Memory: Representation and Remembrance in the Wake of the Holocaust*. Albany: State University of New York Press, 2009.

Bernard-Donals, Michael. "The Rhetoric of Disaster and the Imperative of Writing." *Rhetoric Society Quarterly* 31, no. 1 (2001): 73–94.

Bernard-Donals, Michael, and Kyle Jensen. "Introduction: Taking Rhetoric to Its Limits; or, How to Respond to a Sacred Call." In *Responding to the Sacred: An Inquiry Into the Limits of Rhetoric*, edited by Michael Bernard-Donals and Kyle Jensen, 1–23. University Park: Pennsylvania State University Press, 2021.

Besel, Richard D. "Accommodating Climate Change Science: James Hansen and the Rhetorical/Political Emergence of Global Warming." *Science in Context* 26, no. 1 (2013): 137–52.

Bessette, Joseph A. *The Mild Voice of Reason: Deliberative Democracy and American National Government.* American Politics and Political Economy Series. Chicago: University of Chicago Press, 1997.

Bhagwat, Shonil A., and Claudia Rutte. "Sacred Groves: Potential for Biodiversity Management." *Frontiers in Ecology and the Environment* 4, no. 10 (December 1, 2006): 519–24.

Bhala, Raj. "The Power of the Past: Towards De Jure Stare Decisis in WTO Adjudication (Part Three of a Trilogy)." *George Washington International Law Review* 33, no. 3/4 (2001): 873–978.

Bindé, Jérôme. "Toward an Ethics of the Future." *Public Culture* 12, no. 1 (2000): 51–72.

Björnsson, Helgi. "Glaciers of the Central Highlands." In *The Glaciers of Iceland: A Historical, Cultural and Scientific Overview,* edited by Helgi Björnsson, 275–336. Paris: Atlantis Press, 2017.

Björnsson, Helgi, ed. *The Glaciers of Iceland: A Historical, Cultural and Scientific Overview.* Paris: Atlantis Press, 2017.

Black, Jason Edward. "Here Is a Strange and Bitter Crop: Emmett Till and the Rhetorical Complications of Treescape Memory." *Argumentation and Advocacy* 55, no. 1 (January 2, 2019): 24–41.

Blair, Carole. "Contemporary US Memorial Sites as Exemplars of Rhetoric's Materiality." In *Rhetorical Bodies,* edited by Jack Selzer and Sharon Crowley, 16–57. Madison: University of Wisconsin Press, 1999.

Blair, Carole, Greg Dickinson, and Brian L. Ott. "Introduction: Rhetoric/Memory/Place." In *Places of Public Memory: The Rhetoric of Museums and Memorials,* 1–54. Tuscaloosa: University of Alabama Press, 2010.

Blair, Carole, Marsha S. Jeppeson, and Enrico Pucci. "Public Memorializing in Postmodernity: The Vietnam Veterans Memorial as Prototype." *Quarterly Journal of Speech* 77, no. 3 (August 1, 1991): 263–88.

Blair, Carole, and Neil Michel. "Reproducing Civil Rights Tactics: The Rhetorical Performances of the Civil Rights Memorial." *Rhetoric Society Quarterly* 30, no. 2 (2000): 31–55.

Blair, Carole, V. William Balthrop, and Neil Michel. "Mood of the Material: War Memory and Imagining Otherwise." *Cultural Studies ↔ Critical Methodologies* 13, no. 1 (November 18, 2012): 6–20.

Blommaert, Jan. "Chronotopes, Scales, and Complexity in the Study of Language in Society." *Annual Review of Anthropology* 44, no. 1 (October 21, 2015): 105–16.

Bloome, David, and Laurie Katz. "Literacy as Social Practice and Classroom Chronotopes." *Reading & Writing Quarterly* 13, no. 3 (July 1, 1997): 205–25.

Booth, Wayne C. *The Rhetoric of Fiction.* Literary Criticism. Chicago: University of Chicago Press, 1983.

Booth, Wayne C. *The Rhetoric of Rhetoric: The Quest for Effective Communication.* Malden, MA: Blackwell Publishing, 2004.

Borgmann, Kathi. "Sounds from the Macaulay Library Provide the Backdrop to Maya Lin's Ghost Forest Installation." *The Cornell Lab of Ornithology Macaulay Library* (blog), May 10, 2021. https://www.macaulaylibrary.org/2021/05/10/sounds -from-the-macaulay-library-provide-the-backdrop-to-maya-lins-ghost-forest -installation/.

Boswell, Anna. "Climates of Change: A Tuatara's-Eye View." *Humanities* 9, no. 2 (2020).

Boyer, Dominic. "Planned Remarks." Script for remarks, August 18, 2019. Unpublished manuscript. Shared with author via email by Cymene Howe.

Boykoff, Maxwell. *Creative (Climate) Communications*. Cambridge: Cambridge University Press, 2019.

Boykoff, Maxwell T. *Who Speaks for the Climate? Making Sense of Media Reporting on Climate Change*. Cambridge: Cambridge University Press, 2011.

Boykoff, Maxwell T., and Michael K. Goodman. "Conspicuous Redemption? Reflections on the Promises and Perils of the 'Celebritization' of Climate Change." *Geoforum* 40, no. 3 (May 1, 2009): 395–406.

Boyle, Casey Andrew. *Rhetoric as a Posthuman Practice*. Columbus: Ohio State University Press, 2018.

Boyle, Casey, James J. Brown, and Steph Ceraso. "The Digital: Rhetoric Behind and Beyond the Screen." *Rhetoric Society Quarterly* 48, no. 3 (May 27, 2018): 251–59.

Branswell, Helen. "Why 'Flattening the Curve' May Be the World's Best Bet to Slow the Coronavirus." *Stat*, March 11, 2020. https://www.statnews.com/2020/03/11/flattening-curve-coronavirus/.

Breteau, Pierre. "Coronavirus : Visualisez les pays qui ont «aplati la courbe » de l'épidémie et ceux qui n'y sont pas encore parvenus." *Le Monde*, March 27, 2020.

Broach, Drew. "17 Words, Phrases We Learned in 2020; 'Flatten the Curve' Not a Weight Loss Goal." *Nola.Com*, December 28, 2020. https://www.nola.com/news/article_aff55f70-4189-11eb-807a-2f0a1e6f4e22.html.

Bromfield, Nicholas, and Allan McConnell. "Two Routes to Precarious Success: Australia, New Zealand, COVID-19 and the Politics of Crisis Governance." *International Review of Administrative Sciences* 87, no. 3 (September 1, 2021): 518–35.

Brouwer, Daniel C. "ACT-Ing UP in Congressional Hearings." In *Counterpublics and the State*, edited by Robert Asen and Daniel C. Brouwer, 87–109. Albany: State University of New York Press, 2001.

Bruce, Caitlin Frances. *Painting Publics: Transnational Legal Graffiti Scenes as Spaces for Encounter*. Philadelphia, PA: Temple University Press, 2019.

Brunner, Elizabeth Ann. "Contemporary Environmental Art in China: Portraying Progress, Politics, and Ecosystems." *Environmental Communication* 12, no. 3 (April 3, 2018): 402–13.

Buell, Lawrence. *The Environmental Imagination: Thoreau, Nature Writing, and the Formation of American Culture*. Cambridge, MA: The Belknap Press of Harvard University Press, 1995.

Buell, Lawrence. *Writing for an Endangered World: Literature, Culture, and Environment in the US and Beyond*. Cambridge, MA: The Belknap Press of Harvard University Press, 2001.

Bury, Louis. "Gray and Green: Maya Lin at Madison Square Park." *Art in America*, June 17, 2021. https://www.artnews.com/art-in-america/aia-reviews/maya-lin-madison-square-park-1234596033/.

Byers, Michele S. "Restoring the 'Kidneys of the Pine Barrens.'" *New Jersey Conservation Foundation*, May 30, 2019. https://www.njconservation.org/restoring-the-kidneys-of-the-pine-barrens/.

Byrne, Lydia, Daniel Angus, and Janet Wiles. "Acquired Codes of Meaning in Data Visualization and Infographics: Beyond Perceptual Primitives." *IEEE Transactions on Visualization and Computer Graphics* 22, no. 1 (January 31, 2016): 509–18.

Candeias, Matt. "Trees in a Changing Climate." In Defense of Plants, episode 284. September 27, 2020. https://www.indefenseofplants.com/podcast/tag/forestry.

Carey, Mark. "The History of Ice: How Glaciers Became an Endangered Species." *Environmental History* 12, no. 3 (July 1, 2007): 497–527.

Carey, Tamika L. "Necessary Adjustments: Black Women's Rhetorical Impatience." *Rhetoric Review* 39, no. 3 (July 2, 2020): 269–86.

Centers for Disease Control and Prevention. "Interim Pre-Pandemic Planning Guidance: Community Strategy for Pandemic Influenza Mitigation in the United States—Early, Targeted, Layered Use of Nonpharmaceutical Interventions." Centers for Disease Control and Prevention (US), February 2007. https://stacks.cdc.gov/view/cdc/11425.

Centers for Disease Control and Prevention. "What Is an Epi Curve?" Quick-Learn Lesson, https://www.cdc.gov/training/QuickLearns/createepi/index.html.

Chang, Ailsa. "Interview with Brad Little." *All Things Considered*, National Public Radio, July 20, 2020. https://www.npr.org/2020/07/20/893227053/idaho-governor-on-the-rising-number-of-coronavirus-cases-in-the-state.

Charland, Maurice. "Farrell's Moods." *Philosophy & Rhetoric* 41, no. 4 (2008): 337–55.

Charles, Dan. "US Feels Heat on Global Warming Stance." *New Scientist* 129, no. 1756 (February 16, 1991).

Chirindo, Kundai, Robert Gutierrez-Perez, Matthew Houdek, Louis M. Maraj, Ersula J. Ore, Kendall R. Phillips, Lee M. Pierce, and G. Mitchell Reyes. "Coda: A Rupture in Time." *Women's Studies in Communication*, November 17, 2020, 1–12.

Clary-Lemon, Jennifer. "Gifts, Ancestors, and Relations: Notes Toward an Indigenous New Materialism." *Enculturation: A Journal of Rhetoric, Writing, and Culture*, November 12, 2019. http://enculturation.net/gifts_ancestors_and_relations.

Clary-Lemon, Jennifer. *Planting the Anthropocene: Rhetorics of Natureculture*. Logan: Utah State University Press, 2019.

Cogbill, Charles V., John Burk, and G. Motzkin. "The Forests of Presettlement New England, USA: Spatial and Compositional Patterns Based on Town Proprietor Surveys." *Journal of Biogeography* 29, no. 10/11 (2002): 1279–1304.

Cohen, Jeffrey Jerome. *Stone: An Ecology of the Inhuman*. Minneapolis: University of Minnesota Press, 2015.

Cohen, Jeffrey Jerome, and Lowell Duckert. "Introduction: Eleven Principles of the Elements." In *Elemental Ecocriticism: Thinking with Earth, Air, Water, and Fire*, edited by Jeffrey Jerome Cohen and Lowell Duckert, 2–26. Minneapolis: University of Minnesota Press, 2015.

Congressional Research Service. "Membership of the 116th Congress: A Profile." Report no. R45583, updated December 17, 2020. https://fas.org/sgp/crs/misc/R45583.pdf.

Contini, Carlo, Mariachiara Di Nuzzo, Nicole Barp, Aurora Bonazza, Roberto De Giorgio, Mauro Tognon, and Salvatore Rubino. "The Novel Zoonotic COVID-19 Pandemic: An Expected Global Health Concern." *Journal of Infection in Developing Countries* 14, no. 03 (March 31, 2020): 255–64.

The Cornell Lab. "Are Whip-Poor-Will Populations Declining? What Can We Do About It?" *All About Birds*, April 1, 2009. https://www.allaboutbirds.org/news/are-whip-poor-will-populations-declining-what-can-we-do-about-it/.

Corrigan, Lisa M. *Black Feelings: Race and Affect in the Long Sixties*. Jackson: University Press of Mississippi, 2020.

Coscieme, Luca, Lorenzo Fioramonti, Lars F. Mortensen, Kate E. Pickett, Ida Kubiszewski, Hunter Lovins, Jacqueline Mcglade, et al. "Women in Power: Female Leadership and Public Health Outcomes during the COVID-19 Pandemic." *MedRxiv*, July 16, 2020. https://www.medrxiv.org/content/10.1101/2020.07.13.20152397v2.

Cox, Lisa. "'Unprecedented' Globally: More than 20% of Australia's Forests Burnt in Bushfires." *Guardian*, February 24, 2020. https://www.theguardian.com/australia -news/2020/feb/25/unprecedented-globally-more-than-20-of-australias-forests -burnt-in-bushfires.

Crick, Nathan. "Conquering Our Imagination: Thought Experiments and Enthymemes in Scientific Argument." *Philosophy & Rhetoric* 37, no. 1 (2004): 21–41.

Crutzen, Paul J. "The 'Anthropocene.'" In *Earth System Science in the Anthropocene*, edited by Eckart Ehlers and Thomas Krafft, 13–18. Berlin and Heidelberg: Springer Berlin Heidelberg, 2006.

Crutzen, Paul J., and Eugene F. Stoermer. "The 'Anthropocene.'" *IGBP Newsletter* 41 (2000): 17.

Cunsolo, Ashlee, and Karen Landman. "Introduction: To Mourn Beyond the Human." In *Mourning Nature: Hope at the Heart of Ecological Loss and Grief*, edited by Ashlee Cunsolo and Karen Landman, 3–26. Montreal: McGill-Queen's University Press, 2017.

Davis, Kristine M. "A Description and Analysis of the Legislative Committee Hearing." *Western Journal of Speech Communication* 45, no. 1 (April 1, 1981): 88–106.

DeLaure, Marilyn. "Performing Loss: Sonic Rhetoric in Maya Lin's 'What Is Missing?'" *Liminalities* 16, no. 1 (2020): 1–32.

DeLoughrey, Elizabeth M. *Allegories of the Anthropocene*. Durham, NC: Duke University Press, 2019.

Deloria, Philip. *Indians in Unexpected Places*. Lawrence: University Press of Kansas, 2004.

DeLuca, Kevin Michael. *Image Politics: The New Rhetoric of Environmental Activism*. New York: The Guilford Press, 1999.

DeLuca, Kevin, and Anne Teresa Demo. "Imagining Nature and Erasing Class and Race: Carleton Watkins, John Muir, and the Construction of Wilderness." *Environmental History* 6, no. 4 (2001): 541–60.

Derrida, Jacques. *Sovereignties in Question: The Poetics of Paul Celan*. Edited by Thomas Dutoit and Outi Pasanen. New York: Fordham University Press, 2005.

Detweiler, Eric, and Kate Lisbeth Pantelides. "Predicting Futures, Performing Feminisms: Chronology and Ideology in Composition Classrooms." *Pedagogy* 20, no. 1 (January 1, 2020): 157–70.

DiCaglio, Joshua. *Scale Theory: A Nondisciplinary Inquiry*. Minneapolis: University of Minnesota Press, 2021.

DiCaglio, Joshua, Kathryn M. Barlow, and Joseph S. Johnson. "Rhetorical Recommendations Built on Ecological Experience: A Reassessment of the Challenge of Environmental Communication." *Environmental Communication* 12, no. 4 (May 19, 2018): 438–50.

Dimock, Wai Chee. "Hearing Animals: Thoreau between Fable and Elegy." *J19: The Journal of Nineteenth-Century Americanists* 1, no. 2 (Fall 2013): 397–401.

Dobrin, Sidney I., and Sean Morey. "Mediating Interfaces: Getting In Between." In *Mediating Nature: The Role of Technology in Ecological Literacy*, edited by Sidney I. Dobrin and Sean Morey, 1–14. New York: Routledge, 2020.

Dorman, Tom. "The Year 2020 in Words: The Ones We Can Print." *The Gazette*, December 21, 2020. https://www.thegazette.com/subject/opinion/staff-columnist/the -year-2020-in-words-the-ones-we-can-print-20201221.

Doshi, Rupali Kotwal. "Bending the Curve on Human Immunodeficiency Virus Transmission." *Clinical Infectious Diseases* 71, no. 2 (August 17, 2019): 301–3.

Doss, Erika. *Memorial Mania: Public Feeling in America*. Chicago: University of Chicago Press, 2010.

Douglas, Sarah. "Moderator Remarks." Presented at the Madison Square Conservancy Public Art Symposium: Greening Public Art, New York, June 4, 2021, virtual. https://madisonsquarepark.org/art/videos-and-media/.

Doyle, Julie. "Picturing the Clima(c)Tic: Greenpeace and the Representational Politics of Climate Change Communication." *Science as Culture* 16, no. 2 (June 1, 2007): 129–50.

Doyle, Julie, Nathan Farrell, and Michael K. Goodman. "Celebrities and Climate Change." In *Oxford Encyclopedia of Climate Change Communication*, 95–112. Oxford: Oxford University Press, 2018.

Drouin, Roger. "How Rising Seas Are Killing Southern US Woodlands." *Yale Environment 360*, November 1, 2016. https://e360.yale.edu/features/ghost_forest_rising _sea_levels_killing_coastal_woodlands.

Dryzek, John S., and Alex Y. Lo. "Reason and Rhetoric in Climate Communication." *Environmental Politics* 24, no. 1 (January 2, 2015): 1–16.

Dunmire, Patricia L. "The Rhetoric of Temporality: The Future as Linguistic Concept and Rhetorical Resource." In *Rhetoric in Detail: Discourse Analyses of Rhetorical Talk and Text*, edited by B. Johnstone and C. Eisenhart, 81–111. Discourse Approaches to Politics, Society, and Culture. Amsterdam and Philadelphia: John Benjamins, 2008.

Dyson, Francis. *Sounding New Media: Immersion and Embodiment in the Arts and Culture*. Berkeley: University of California Press, 2009.

Duxbury, Neil. *The Nature and Authority of Precedent*. Cambridge: Cambridge University Press, 2008.

Einarsson, T., and K. J. Albertsson. "The Glacial History of Iceland During the Past Three Million Years." *Philosophical Transactions of the Royal Society of London. Series B, Biological Sciences* 318, no. 1191 (1988).

Ellis, John. *Seeing Things: Television in the Age of Uncertainty*. London: I. B. Tauris Publishers, 2000.

Engel, Currie. "Scientists Unveil Memorial to Iceland's 'First' Dead Glacier Lost to Climate Change." *Time*, July 22, 2019. https://time.com/5631599/iceland-glacier -climate-change/.

Entrèves, Maurizio Passerin d'. *The Political Philosophy of Hannah Arendt*. New York: Routledge, 1994.

Estrada-Peña, Agustin, Richard S. Ostfeld, A. Townsend Peterson, Robert Poulin, and José de la Fuente. "Effects of Environmental Change on Zoonotic Disease Risk: An Ecological Primer." *Trends in Parasitology* 30, no. 4 (April 2014): 205–14.

Ewalt, Joshua P. "Mapping Injustice: The World Is Witness, Place-Framing, and the Politics of Viewing on Google Earth." *Communication, Culture and Critique* 4, no. 4 (December 1, 2011): 333–54.

Facourt, Stéphane. "Le taux de CO2 dans l'air au plus haut depuis plus de 2,5 millions d'années." *Le Monde*, May 6, 2013.

Fahnestock, Jeanne. "The New Mathematical Arts of Argument: Naturalistic Images and Geometric Diagrams." In *Arguing with Numbers: The Intersections of Rhetoric and Mathematics*, edited by James Wynn and G. Mitchell Reyes, 171–211. University Park: Pennsylvania State University Press, 2021.

Fahnestock, Jeanne. *Rhetorical Figures in Science*. Rhetorical Figures in Science. Oxford: Oxford University Press, 1999.

Fahnestock, Jeanne. *Rhetorical Style*. New York: Oxford University Press, 2011.

Farhadi, Hamed, Mohsen Faizi, and Hanyieh Sanaieian. "Mitigating the Urban Heat Island in a Residential Area in Tehran: Investigating the Role of Vegetation, Materials, and Orientation of Buildings." *Sustainable Cities and Society* 46 (April 1, 2019): 101448.

Farrell, Thomas B. "Sizing Things Up: Colloquial Reflection as Practical Wisdom." *Argumentation* 12 (1998): 1–14.

Farrell, Thomas B. "The Weight of Rhetoric: Studies in Cultural Delirium." *Philosophy & Rhetoric* 41, no. 4 (2008): 467–87.

Felman, Shoshana. "Fire in the Archive: The Alignment of Witnesses." In *The Future of Testimony*, 48–68. New York: Routledge, 2014.

Felsenthal, Edward. "Person of the Year: The Choice." *Time*, December 23, 2019.

Fifield, Anna. "New Zealand Isn't Just Flattening the Curve: It's Squashing It." *Washington Post*, April 7, 2020, sec. World. https://www.washingtonpost.com/world/asia_pacific/new-zealand-isnt-just-flattening-the-curve-its-squashing-it/2020/04/07/6cab3a4a-7822-11ea-a311-adb1344719a9_story.html.

Fihn, Stephan D. "Bending the Curve on Cardiovascular Risk." *JAMA Internal Medicine* 174, no. 1 (January 1, 2014): 48–50.

Finnegan, Cara. *Making Photography Matter*. Urbana: University of Illinois Press, 2015.

"Flattening the Other Curve." *The Economist* 435.9195, May 23, 2020, 14–15.

Fleckenstein, Kristie. "The Nineteenth-Century Parlor Stereoscope and the Second-Naturing of Vision." In *Rhetoric, Through Everyday Things*, edited by Scot Barnett and Casey Boyle, 125–38. Tuscaloosa: University of Alabama Press, 2016.

Fleming, James Rodger. "Picturing Climate Control: Visualizing the Unimaginable." In *Image Politics of Climate Change*, 345–62. Bielefeld: Transcript-Verlag, 2014.

Flowers, Katherine S. "Making English Local: Chronotopes in Language Policy Discourse." *Journal of Language and Politics* 15, no. 4 (2016): 469–92.

Foer, Franklin. "Greta Thunberg Is Right to Panic." *Atlantic*, September 20, 2019. https://www.theatlantic.com/ideas/archive/2019/09/greta-thunbergs-despair-is-entirely-warranted/598492/.

Foust, Christina R., and William O'Shannon Murphy. "Revealing and Reframing Apocalyptic Tragedy in Global Warming Discourse." *Environmental Communication* 3, no. 2 (July 1, 2009): 151–67.

France Insoumise Montréal. "Aplatissons Aussi Cette Courbe-Là." Graphic attached to "Une fois qu'on aura réussi à aplatir la courbe du Covid-19, il ne faudra pas oublier d'aplatir celle des changements climatiques!" Facebook, April 9, 2020. https://www.facebook.com/MontrealFI/photos/a.1137163443095808/2079447012200775.

Friedman, Thomas L. "Global Weirding Is Here." *New York Times*, February 17, 2010. https://www.nytimes.com/2010/02/17/opinion/17friedman.html.

Friedman, Thomas L. "The People We Have Been Waiting For." *New York Times*, December 2, 2007. https://www.nytimes.com/2007/12/02/opinion/02friedman.html.

Frost, Robert. *A Witness Tree*. New York: Henry Holt and Company, 1942.

Fukushima, Annie Isabel. *Migrant Crossings: Witnessing Human Trafficking in the US*. Stanford, CA: Stanford University Press, 2019.

"Full Metre Nine Thy Father Lies." *Guardian*, April 26, 1968, City Edition, 1.

Fuller, Thomas, and Mike Baker. "Coronavirus Death in California Came Weeks Before First Known US Death." *New York Times*, April 22, 2020. https://www.nytimes.com/2020/04/22/us/coronavirus-first-united-states-death.html.

Gallagher, Michael. "Field Recording and the Sounding of Spaces." *Environment and Planning D: Society and Space* 33, no. 3 (June 1, 2015): 560–76.

Galloway, George Barnes, and United States Congress. House. Committee on House Administration. *History of the United States House of Representatives.* 87th Congress, 1st Session. House Document. Washington, DC: US Government Printing Office, 1962.

Geertz, Clifford. *The Interpretation of Cultures.* New York: Basic Books, 1973.

Ghosh, Amitav. *The Great Derangement: Climate Change and the Unthinkable.* Chicago: University of Chicago Press, 2018.

"Ghost Forest—Madison Square Park: 10,000 BCE-2020 CE." *What Is Missing? A Memorial.* What Is Missing Foundation. https://whatismissing.org/timeline /ghostforest. Accessed July 15, 2022.

Givoni, Michal. *The Care of the Witness: A Contemporary History of Testimony in Crises.* Cambridge: Cambridge University Press, 2016.

Glenn, Cheryl. *Rhetorical Feminism and This Thing Called Hope.* Carbondale: Southern Illinois University Press, 2018.

Goodman, Jennifer. "$2.3B Office Tower Breaks Ground in New York City Despite COVID-19 Challenges." *Construction Dive,* November 18, 2020. https://www .constructiondive.com/news/23b-office-tower-breaks-ground-in-new-york-city -despite-covid-19-challeng/589274/.

Goodnight, G. Thomas. "The Personal, Technical, and Public Spheres of Argument: A Speculative Inquiry into the Art of Public Deliberation." *Journal of the American Forensic Association* 18, no. 4 (March 1, 1982): 214–27.

Goodnight, G. Thomas. "Public Discourse." *Critical Studies in Mass Communication* 4, no. 4 (December 1, 1987): 428–32.

Gordon, Avery F. *Ghostly Matters: Haunting and the Sociological Imagination.* 2nd edition. Minneapolis: University of Minnesota Press, 2008.

Gordon, Avery F. "Some Thoughts on Haunting and Futurity." *Borderlands* 10, no. 2 (2011): 1–21.

Gordon, Jeremy G., Katherine D. Lind, and Saul Kutnicki. "A Rhetorical Bestiary." *Rhetoric Society Quarterly* 47, no. 3 (May 27, 2017): 222–28.

Gore, Albert. "Remarks by Vice President Albert Gore." Glacier National Park, September 2, 1997. https://clintonwhitehouse2.archives.gov/WH/EOP/OVP /speeches/glacier.html.

Gounder, Celine. "A Second Wave?" *Epidemic,* Season 1, June 19, 2020. https://www .justhumanproductions.org/epidemic-s1e30-transcript.

Government of Canada. "2 Billion Tree Commitment." Accessed July 20, 2021. https:// www.canada.ca/en/campaign/2-billion-trees.html.

Gries, Laurie A. *Still Life with Rhetoric: A New Materialist Approach for Visual Rhetorics.* Logan: Utah State University Press, 2015.

Grieve, Sarah. "Environmental Justice Witnessing in Muriel Rukeyser's *The Book of the Dead.*" *ISLE: Interdisciplinary Studies in Literature and Environment* 26, no. 4 (November 1, 2019): 968–85.

Grignon, Jeff, and Robin Wall Kimmerer. "Listening to the Forest." In *Wildness: Relations of People and Place,* edited by Gavin Van Horn and John Hausdoerffer, 67–74. Chicago: University of Chicago Press, 2017.

Gronbeck, Bruce E. "Rhetorical Timing in Public Communication." *Central States Speech Journal* 25, no. 2 (June 1, 1974): 84–94.

Grosz, Elizabeth A. *Chaos, Territory, Art: Deleuze and the Framing of the Earth.* The Welleck Library Lectures. New York: Columbia University Press, 2008.

Grosz, P. G. *On the Grammar of Optative Constructions.* Linguistik Aktuell/Linguistics Today. Amsterdam and Philadelphia: John Benjamins Publishing Company, 2012.

Gruen, Nicholas. "The Curve We're Not Flattening." Twitter Post, @ngruen1, March 25, 2020. https://twitter.com/NGruen1/status/1242981720191684608.

Gunn, Joshua. "Mourning Speech: Haunting and the Spectral Voices of Nine-Eleven." *Text and Performance Quarterly* 24, no. 2 (April 1, 2004): 91–114.

Gunn, Joshua. "Review Essay: Mourning Humanism; or, The Idiom of Haunting." *Quarterly Journal of Speech* 92, no. 1 (February 1, 2006): 77–102.

Gunn, Joshua. "Speech's Sanatorium." *Quarterly Journal of Speech* 101, no. 1 (January 2, 2015): 18–33.

Guterres, António. "Statement by the Secretary-General on the IPCC Special Report Global Warming of 1.5°C." United Nations, Secretary-General, New York, October 8, 2018. https://www.un.org/sg/en/content/sg/statement/2018-10-08/statement-secretary-general-ipcc-special-report-global-warming-15-%C2%BAc.

Halloran, Michael. "Doing Public Business in Public." In *Form and Genre: Shaping Rhetorical Action,* edited by Karlyn Kohrs Campbell and Kathleen Hall Jamieson, 118–38. Falls Church, VA: Speech Communication Association, 1978.

Hansen, James, and Sato Makiko. "Greenhouse Gas Growth Rates." *Proceedings of the National Academy of Sciences* 101, no. 46 (November 16, 2004): 16109–14.

Hansen, Kathryn. "Okjökull Remembered." NASA Earth Observatory, August 9, 2019. https://earthobservatory.nasa.gov/images/145439/okjokull-remembered.

Hariman, Robert, and John Louis Lucaites. "Icons, Appropriations, and the Co-Production of Meaning." In *Rhetorical Audience Studies and Reception of Rhetoric,* edited by Jens E. Kjeldsen, 285–308. Cham, Switzerland: Palgrave Macmillan, 2018.

Harris, Drew. "Flattening the Coronavirus Curve Goes Way Beyond Science." *Philadelphia Inquirer,* March 24, 2020. https://www.inquirer.com/health/coronavirus/coronavirus-covid-19-drew-harris-infographic-20200324.html.

Harris, Drew. "Important to Remember that #COVID-19 Epidemic Control Measures May Only Delay Cases, Not Prevent." Twitter post, @drewaharris, February 28, 2020. https://twitter.com/drewaharris/status/1233267475036372992?lang=en.

Hawhee, Debra. *Bodily Arts: Rhetoric and Athletics in Ancient Greece.* Austin: University of Texas Press, 2004.

Hawhee, Debra. "Kairotic Encounters." In *Perspectives on Rhetorical Invention,* 16–35. Knoxville: University of Tennessee Press, 2003.

Hawhee, Debra. "Looking into Aristotle's Eyes: Toward a Theory of Rhetorical Vision." *Advances in the History of Rhetoric* 14 (2011): 139–65.

Hawhee, Debra. *Rhetoric in Tooth and Claw: Animals, Language, Sensation.* Chicago: University of Chicago Press, 2017.

Hawhee, Debra. "Rhetoric's Sensorium." *Quarterly Journal of Speech* 101, no. 1 (2015): 2–17.

Hawhee, Debra, and Megan Poole. "Kenneth Burke at the MoMA: A Viewer's Theory." *Quarterly Journal of Speech* 105, no. 4 (October 2, 2019): 418–40.

Hawk, Byron. *Resounding the Rhetorical: Composition as a Quasi-Object.* Composition, Literacy, and Culture. Pittsburgh, PA: University of Pittsburgh Press, 2018.

Hayes, Christal. "'I Want You to Take Action': Greta Thunberg, Teenage Climate Activist, Testifies Before Congress." *USA Today,* September 18, 2019.

Hayes, David J. "Flatten the (Climate) Curve." *State Energy & Environmental Impact Center, NYU School of Law* (blog). Accessed June 23, 2020. https://www.law.nyu

.edu/centers/state-impact/press-publications/expert-commentary/flatten-climate
-curve.

Hayhoe, Katharine. "What Is Global Weirding?" Global Weirding with Katharine Hayhoe. YouTube video, 1:00, September 13, 2016. https://www.youtube.com/watch?v=yc3rYFcAyeI.

Hayman, Eleanor, Colleen James, and Mark Wedge. "Future Rivers of the Anthropocene or Whose Anthropocene Is It? Decolonising the Anthropocene!" *Decolonization: Indigeneity, Education & Society* 6, no. 2 (2018): 77–92.

Heise, Ursula K. *Imagining Extinction*. Chicago: University of Chicago Press, 2016.

Heise, Ursula K. *Sense of Place and Sense of Planet: The Environmental Imagination of the Global*. Oxford: Oxford University Press, 2008.

Helfand, Zach. "Ghost Forest." *New Yorker*, May 17, 2021, 18–19.

Helgason, T., T. J. Daniell, R. Husband, A. H. Fitter, and J. P. W. Young. "Ploughing Up the Wood-Wide Web?" *Nature* 394, no. 6692 (July 1, 1998): 431.

Herrera, Tim. "Left Speechless? Let's Review." *New York Times*, December 20, 2020, Sunday Styles section.

Hmiel, Benjamin, V. V. Petrenko, M. N. Dyonisius, C. Buizert, A. M. Smith, P. F. Place, C. Harth, et al. "Preindustrial 14CH4 Indicates Greater Anthropogenic Fossil CH4 Emissions." *Nature* 578, no. 7795 (February 1, 2020): 409–12.

Hoag, Trevor. "Ghosts of Memory: Mournful Performance and the Rhetorical Event of Haunting (Or: Specters of Occupy)." *Liminalities: A Journal of Performance Studies* 10, no. 3/4 (2014): 1–22.

Hogg, James, Maria Fonoberova, and Igor Mezić. "Exponentially Decaying Modes and Long-Term Prediction of Sea Ice Concentration Using Koopman Mode Decomposition." *Scientific Reports* 10, no. 1 (October 1, 2020): 16313.

Holshue, Michelle L., Chas DeBolt, Scott Lindquist, Kathy H. Lofy, John Wiesman, Hollianne Bruce, Christopher Spitters, et al. "First Case of 2019 Novel Coronavirus in the United States." *New England Journal of Medicine* 382, no. 10 (January 31, 2020): 929–36.

Holthaus, Eric. "This is the first time in human history our planet's atmosphere has had more than 415 ppm CO$_2$. Not just in recorded history, not just since the invention of agriculture 10,0000 years ago. Since before modern humans existed millions of years ago. We don't know a planet like this." Twitter post, May 12, 2019. https://twitter.com/ericholthaus/status/1127681719216353280.

Horowitz, Jason. "Athens Is Only Getting Hotter. Its New 'Chief Heat Officer' Hopes to Cool It Down." *New York Times*, August 21, 2021. https://www.nytimes.com/2021/08/21/world/europe/athens-is-only-getting-hotter-its-new-chief-heat-officer-hopes-to-cool-it-down.html.

Horton, Adrian. "'Odd, Eerie and Haunting': Behind Maya Lin's Manhattan Ghost Forest." *Guardian*, May 13, 2021. https://www.theguardian.com/artanddesign/2021/may/13/maya-lin-manhattan-madison-square-park.

Houdek, Matthew, and Kendall Phillips. "Rhetoric and the Temporal Turn: Race, Gender, Temporalities." *Women's Studies in Communication*, December 4, 2020, 369–83.

Houser, Heather. *Infowhelm: Environmental Art and Literature in an Age of Data*. New York: Columbia University Press, 2020.

Howe, Cymene. "Planned Remarks." Script for remarks, August 18, 2019. Unpublished manuscript shared with author via email.

Howe, Cymene. Virtual interview by author, State College, PA, and Berlin, Germany, October 10, 2019.

Howe, Cymene, and Dominic Boyer. *Not OK: A Little Movie about a Small Glacier at the End of the World*. Documentary. Mingomena, 2018. https://notokmovie.com.

Howe, Cymene, and Dominic Boyer. Unpublished proposal to Ragnar Frank, November 11, 2018.

Hunt, Everett. "The Rhetorical Mood of World War II." *Quarterly Journal of Speech* 29, no. 1 (February 1, 1943): 1–5.

I Am Greta. Directed by Nathan Grossman. Stockholm, Sweden: B-Reel Films, 2020.

Intergovernmental Panel on Climate Change. *Climate Change 2021: The Physical Science Basis*. Contribution of Working Group I to the Sixth Assessment Report of the Intergovernmental Panel on Climate Change, edited by V. Masson-Delmotte, P. Zhai, A. Pirani, S. L. Connors, C. Péan, S. Berger, N. Caud, Y. Chen, L. Goldfarb, M. I. Gomis, M. Huang, K. Leitzell, E. Lonnoy, J. B. R. Matthews, T. K. Maycock, T. Waterfield, O. Yelekçi, R. Yu, and B. Zhou. Cambridge: Cambridge University Press, 2021.

Intergovernmental Panel on Climate Change. "Climate Change Widespread, Rapid, and Intensifying—IPCC." August 9, 2021. https://www.ipcc.ch/2021/08/09/ar6-wg1-20210809-pr/.

Intergovernmental Panel on Climate Change. *IPCC 2018: Global Warming of 1.5°C*. An IPCC Special Report on the impacts of global warming of 1.5°C above pre-industrial levels and related global greenhouse gas emission pathways, in the context of strengthening the global response to the threat of climate change, sustainable development, and efforts to eradicate poverty. Edited by V. Masson-Delmotte, P. Zhai, H.-O. Pörtner, D. Roberts, J. Skea, P. R. Shukla, A. Pirani, W. Moufouma-Okia, C. Péan, R. Pidcock, S. Connors, J. B. R. Matthews, Y. Chen, X. Zhou, M. I. Gomis, E. Lonnoy, T. Maycock, M. Tignor, and T. Waterfield. Geneva, Switzerland: IPCC, 2019.

Jack, Jordynn. "Chronotopes: Forms of Time in Rhetorical Argument." *College English* 69, no. 1 (2006): 52–73.

Jack, Jordynn. "A Pedagogy of Sight: Microscopic Vision in Robert Hooke's Micrographia." *Quarterly Journal of Speech* 95, no. 2 (May 2009): 192–209.

Jackson, M. *The Secret Lives of Glaciers*. West Brattleboro, VT: Green Writers Press, 2019.

Jakobsdóttir, Katrín. "Iceland's Prime Minister: 'The Ice Is Leaving.'" *New York Times*, August 17, 2019, Sunday Review section. https://www.nytimes.com/2019/08/17/opinion/iceland-glacier-climate-change.html.

Jamieson, Thomas. "'Go Hard, Go Early': Preliminary Lessons from New Zealand's Response to COVID-19." *American Review of Public Administration* 50, no. 6–7 (August 1, 2020): 598–605.

Johnson, Elizabeth, Harlan Morehouse, Simon Dalby, Jessi Lehman, Sara Nelson, Rory Rowan, Stephanie Wakefield, and Kathryn Yusoff. "After the Anthropocene: Politics and Geographic Inquiry for a New Epoch." *Progress in Human Geography* 38, no. 3 (June 1, 2014): 439–56.

Johnson, Jenell. "'A Man's Mouth Is His Castle': The Midcentury Fluoridation Controversy and the Visceral Public." *Quarterly Journal of Speech* 102, no. 1 (January 2, 2016): 1–20.

Johnson, Lacy. "How to Mourn a Glacier." *New Yorker*, October 20, 2019. https://www.newyorker.com/news/dispatch/how-to-mourn-a-glacier.

Johnson, Laura. "(Environmental) Rhetorics of Tempered Apocalypticism in *An Inconvenient Truth*." *Rhetoric Review* 28, no. 1 (January 2, 2009): 29–46.

Johnstone, Henry. "Rhetoric as a Wedge: A Reformulation." *Rhetoric Society Quarterly* 20, no. 4 (1990): 333–38.

Jones, Madison. "Sylvan Rhetorics: Roots and Branches of More-than-Human Publics." *Rhetoric Review* 38, no. 1 (January 2, 2019): 63–78.

Jotzo, Frank, Joanna Depledge, and Harald Winkler. "US and International Climate Policy under President Trump." *Climate Policy* 18, no. 7 (August 9, 2018): 813–17.

Kahneman, Daniel. *Thinking, Fast and Slow.* New York: Farrar, Straus and Giroux, 2011.

Keller, Lynn. *Recomposing Ecopoetics: North American Poetry of the Self-Conscious Anthropocene.* Charlottesville: University of Virginia Press, 2017.

Kennerly, Michele. *Editorial Bodies: Perfection and Rejection in Ancient Rhetoric and Poetics.* Columbia: University of South Carolina Press, 2018.

Kennerly, Michele. "Getting Carried Away: How Rhetorical Transport Gets Judgment Going." *Rhetoric Society Quarterly* 40, no. 3 (2010): 269–91.

Khan, M. L., Ashalata Devi Khumbongmayum, and R. S. Tripathi. "The Sacred Groves and Their Significance in Conserving Biodiversity: An Overview." *International Journal of Ecology and Environmental Sciences* 34, no. 3 (2008): 277–91.

Kimmerer, Robin Wall. *Braiding Sweetgrass: Indigenous Wisdom, Scientific Knowledge and the Teachings of Plants.* Minneapolis, MN: Milkweed Editions, 2013.

King, Martin Luther, Jr. "I Have a Dream." Lincoln Memorial, Washington, DC. August 28, 1963. Transcript at *American Rhetoric.* https://www.americanrhetoric.com/speeches/mlkihaveadream.htm, accessed July 15, 2022.

Kirwan, Matthew L., and Keryn B. Gedan. "Sea-Level Driven Land Conversion and the Formation of Ghost Forests." *Nature Climate Change* 9, no. 6 (June 1, 2019): 450–57.

Kolbert, Elizabeth. "How Iceland Beat the Coronavirus." *New Yorker*, June 1, 2020. https://www.newyorker.com/magazine/2020/06/08/how-iceland-beat-the-coronavirus.

Kolodziejski, Lauren R. "What Is Missing? Reflections on the Human–Nature Relationship in Maya Lin's Final Memorial." *Environmental Communication* 9, no. 4 (October 2, 2015): 428–45.

Kostelnick, Charles. "The Visual Rhetoric of Data Displays: The Conundrum of Clarity." *IEEE Transactions on Professional Communication* 51, no. 1 (March 2008): 116–30.

Kostelnick, Charles, and Michael Hassett. *Shaping Information: The Rhetoric of Visual Conventions.* Carbondale: Southern Illinois University Press, 2003.

Kummer, Frank. "Rising Seas Could Be Turning Jersey's Coastal Cedars into Ghost Forests." *Philadelphia Inquirer*, May 21, 2017. https://www.inquirer.com/philly/health/environment/new-jerseys-atlantic-white-cedar-ghost-forests-harbingers-of-sea-level-rise-rutgers-20170521.html.

Kunreuther, Howard, and Paul Slovic. "What the Coronavirus Curve Teaches Us About Climate Change." *Politico*, March 26, 2020. https://www.politico.com/news/magazine/2020/03/26/what-the-coronavirus-curve-teaches-us-about-climate-change-148318.

Kyff, Rob. "Word Watch: Flatten the Curve, Ahead of It, and a Steep Learning Curve." *Hartford Courant*, March 28, 2020. https://www.courant.com/ctnow/arts-theater/hc-ctnow-word-watch-0329-20200328-qge3bglmfjbkda2wply6sdfzty-story.html.

LaBelle, Brandon. *Acoustic Justice: Listening, Performativity, and the Work of Reorientation.* London: Bloomsbury Academic, 2021.

Lai, K. K. Rebecca, and Keith Collins. "Which Country Has Flattened the Curve for the Coronavirus?" *New York Times*, March 19, 2020, World section. https://www

.nytimes.com/interactive/2020/03/19/world/coronavirus-flatten-the-curve
-countries.html.

Lam, Anita, and Matthew Tegelberg. "Witnessing Glaciers Melt: Climate Change and Transmedia Storytelling." *Journal of Science Communication* 18, no. 2 (2019): A05.

Lammers, Joris, Jan Crusius, and Anne Gast. "Correcting Misperceptions of Exponential Coronavirus Growth Increases Support for Social Distancing." *Proceedings of the National Academy of Sciences* 117, no. 28 (2020): 16264–66.

Larson, Stephanie R. "'Just Let This Sink in': Feminist Megethos and the Role of Lists in #MeToo." *Rhetoric Review* 38, no. 4 (October 2, 2019): 432–44.

Larson, Stephanie R. *What It Feels Like: Visceral Rhetoric and the Politics of Rape Culture*. University Park: Pennsylvania State University Press, 2021.

"The Last Heath Hen." *Scientific Monthly* 32, no. 4 (1931): 382–84.

"The Last of Ice: Okjökull." *The Economist*, September 21, 2019.

Latour, Bruno. *Facing Gaia: Eight Lectures on the New Climatic Regime*. Cambridge: Polity Press, 2017.

Latour, Bruno. "Waiting for Gaia. Composing the Common World through Arts and Politics." Lecture delivered at the French Institute, London, November 2011. http://www.bruno-latour.fr/sites/default/files/124-GAIA-LONDON-SPEAP_0.pdf.

Latour, Bruno. "Why Has Critique Run Out of Steam? From Matters of Fact to Matters of Concern." *Critical Inquiry* 30, no. 2 (January 1, 2004): 225–48.

LaWare, Margaret, and Victoria J. Gallagher. "The Power of Agency: Urban Communication and the Rhetoric of Public Art." In *The Urban Communication Reader*, edited by Gene Burd, Susan Drucker, and Gary Gumpert, 161–73. New York: Hampton Press, 2007.

Leiserowitz, Anthony, and Nicholas Smith. "Affective Imagery, Risk Perception, and Climate Change Communication." In *Oxford Research Encyclopedia of Climate Science*, 3. Oxford: Oxford University Press, 2017.

Leiserowitz, Anthony, and Nicholas Smith. "The Rise of Global Warming Skepticism: Exploring Affective Image Associations in the United States Over Time." *Risk Analysis* 32, no. 6 (2012): 1021–32.

Lemon, Alaina. "Sympathy for the Weary State? Cold War Chronotopes and Moscow Others." *Comparative Studies in Society and History* 51, no. 4 (2009): 832–64.

Le Quéré, Corinne, Robert B. Jackson, Matthew W. Jones, Adam J. P. Smith, Sam Abernethy, Robbie M. Andrew, Anthony J. De-Gol, et al. "Temporary Reduction in Daily Global CO_2 Emissions during the COVID-19 Forced Confinement." *Nature Climate Change* 10, no. 7 (July 1, 2020): 647–53.

"Let's Flatten This Curve Too." Instagram post, @thesustainablefashionforum, March 27, 2020. https://www.instagram.com/p/B-PxHoVHoFE/?hl=en.

Leveille, David. "Artist Maya Lin's Sound Ring Immerses Listeners in a World of Endangered Species." *The World*, Public Radio International, June 12, 2014. https://www.pri.org/stories/2014-06-12/artist-maya-lins-sound-ring-immerses-listeners
-world-endangered-species.

Levine, Michael. *The Belated Witness: Literature, Testimony, and the Question of Holocaust Survival*. Stanford, CA: Stanford University Press, 2006.

Levy, Matthew R., and Joshua Tasoff. "Exponential-Growth Bias and Overconfidence." *Journal of Economic Psychology* 58 (2017): 1–14.

Lewis, Charleton T., and Charles Short. *A Latin Dictionary, Founded on Andrews' Edition of Freund's Latin Dictionary*. Oxford: Clarendon Press, 1980.

Lewis, Simon L., and Mark A. Maslin. "Defining the Anthropocene." *Nature* 519 (March 12, 2015): 171–80.

Liddell, Henry George, Robert Scott, and Sir Henry Stuart Jones. *Greek-English Lexicon.* 9th edition. Oxford: Oxford University Press, 1996.

Lin, Maya. "Press Preview Remarks." Madison Square Park and Madison Square Park Conservancy, New York, New York, May 10, 2021. Transcribed by the author.

Lukpat, Alyssa, and Sarah Cahalan. "The US Breaks Its Single-Day Case Record, Nearly Doubling the Highest Numbers from Last Winter." *New York Times,* December 30, 2021. https://www.nytimes.com/2021/12/30/world/the-us-breaks-its-single-day-case-record-nearly-doubling-the-highest-numbers-from-last-winter.html.

Lynch, John. "Generational Chronotopes and Accounting for Unethical Medicine: Bill Clinton's Apologies for Radiation Research and the Tuskegee Syphilis Study." *Quarterly Journal of Speech* 107, no. 3 (2021): 284–304.

Lyons, James. "'Gore Is the World': Embodying Environmental Risk in *An Inconvenient Truth.*" *Journal of Risk Research* 22, no. 9 (September 2, 2019): 1156–70.

Lyons, Maura. "An Embodied Landscape: Wounded Trees at Gettysburg." *American Art* 26, no. 3 (September 1, 2012): 44–65.

MacCarthy, J. J., J. T. Houghton, R. T. Watson, Groupe d'experts intergouvernemental sur l'évolution du climat, Intergovernmental Panel on Climate Change, D. L. Albritton, Core Writing Team, Science Intergovernmental Panel on Climate Change. Working Group 1, Impacts Intergovernmental Panel on Climate Change. Working Group 2, and Cambridge University Press. *Climate Change 2001: Synthesis Report: Third Assessment Report of the Intergovernmental Panel on Climate Change.* Climate Change 2001. Cambridge: Cambridge University Press, 2001.

Mace, Georgina M., Mike Barrett, Neil D. Burgess, Sarah E. Cornell, Robin Freeman, Monique Grooten, and Andy Purvis. "Aiming Higher to Bend the Curve of Biodiversity Loss." *Nature Sustainability* 1, no. 9 (September 1, 2018): 448–51.

Mackenzie, John S., and David W. Smith. "COVID-19: A Novel Zoonotic Disease Caused by a Coronavirus from China: What We Know and What We Don't." *Microbiology Australia* 41, no. 1 (March 17, 2020): 45–50.

Mackenzie, John S., and Martyn Jeggo. "Reservoirs and Vectors of Emerging Viruses." *Current Opinion in Virology* 3, no. 2 (2013): 170–79.

Madison Square Park Conservancy. "Garden Jewels, April 12–30." Past Exhibits. https://madisonsquarepark.org/horticulture/exhibitions/view/garden-jewels/.

Madison Square Park Conservancy. "Ghost Forest Full Documentary." YouTube video, 4:16, May 10, 2021. https://www.youtube.com/watch?v=0DHpGPwHm2I.

Madison Square Park Conservancy. "Maya Lin Conceives a Soundscape for *Ghost Forest.*" Community News, May 25, 2021. https://madisonsquarepark.org/community/news/2021/05/maya-lin-conceives-a-soundscape-for-ghost-forest/.

Madison Square Park Conservancy. "Maya Lin *Ghost Forest* Soundscape," May 10, 2021. Audio available at https://madisonsquarepark.org/art/exhibitions/maya-lin-ghost-forest/.

Madison Square Park Conservancy. "Press Release: Maya Lin Ghost Forest." April 1, 2021; last updated May 10, 2021. https://18p95aamd693k2glx3ytryxr-wpengine.netdna-ssl.com/wp-content/uploads/2021/02/Maya-Lin-Ghost-Forest-Press-Release_FINAL.pdf.

Madison Square Park Conservancy. "Step into Our Tree Museum." Horticulture, Arboretum, accessed May 10, 2021. https://madisonsquarepark.org/horticulture/arboretum/.

Magnason, Andri Snær. "The Glaciers of Iceland Seemed Eternal. Now a Country Mourns Their Loss." *Guardian*, August 14, 2019. https://www.theguardian.com /commentisfree/2019/aug/14/glaciers-iceland-country-loss-plaque-climate -crisis.

Magnason, Andri Snær. "A Letter to the Future." Plaque placed on Ok glacier, August 18, 2019, Ok, Borgarnes, Iceland.

Makower, Joel. "Can We Flatten the Curve on Climate?" *GreenBiz*, April 7, 2020. https://www.greenbiz.com/article/can-we-flatten-curve-climate.

Mandel, Emily St. John. *Station Eleven*. London: Pan Macmillan, 2014.

Mandhana, Niharika. "Some Countries Are Squashing the Coronavirus Curve. Vietnam Is One." *Wall Street Journal*, April 27, 2020. https://www.wsj.com/articles/some -countries-are-squashing-the-coronavirus-curve-vietnam-is-one-11587989361.

Mann, Michael E. *The New Climate War: The Fight to Take Back Our Planet*. New York: Public Affairs, 2021.

Manning, Erin. *The Minor Gesture*. Thought in the Act. Durham, NC: Duke University Press, 2016.

Mapes, Lynda V. *Witness Tree: Seasons of Change with a Century-Old Oak*. London: Bloomsbury Publishing, 2017.

Marcott, Shaun A., Jeremy D. Shakun, Peter U. Clark, and Alan C. Mix. "A Reconstruction of Regional and Global Temperature for the Past 11,300 Years." *Science* 339, no. 6124 (March 8, 2013): 1198.

Margolin, Jamie. *Youth to Power: Your Voice and How to Use It*. Paris: Hachette Books, 2020.

Markel, Howard. "What History Revealed About Cities that Socially Distanced During a Pandemic." *PBS News Hour*, April 20, 2020. https://www.pbs.org/newshour /health/what-history-can-teach-us-about-flattening-the-curve.

Markel, Howard, Harvey B. Lipman, J. Alexander Navarro, Alexandra Sloan, Joseph R. Michalsen, Alexandra Minna Stern, and Martin S. Cetron. "Nonpharmaceutical Interventions Implemented by US Cities During the 1918–1919 Influenza Pandemic." *JAMA* 298, no. 6 (August 8, 2007): 644–54.

Martellozzo, Nicola. "Wind, Wood, and the Entangled Life of Disasters." *HAU: Journal of Ethnographic Theory* 11, no. 2 (September 1, 2021).

Martyr, Pietro d'Anghiera. *The Decades of the Newe Worlde Or West India* Translated by Richard Eden. Early English Books Online/EEBO. London: In ædibus Guilhelmi Powell [for Robert Toy], 1555.

Maxson, David. "Rhetorical Burials: Memorial Practices of New Orleans." PhD diss., The Pennsylvania State University, 2018.

"Maya Lin *Ghost Forest*." Placard, *Ghost Forest*, Madison Square Park, May 10– November 14, 2021.

McGreevy, Ronan. "Covid-19, Cocooning, New Normal—10 Phrases that Defined 2020." *Irish Times*, December 31, 2020. https://www.irishtimes.com/news/ireland /irish-news/covid-19-cocooning-new-normal-10-phrases-that-defined-2020-1 .4448012.

McKibben, Bill. "130 Degrees." *New York Review of Books*, August 20, 2020. https:// www.nybooks.com/articles/2020/08/20/climate-emergency-130-degrees/.

McPhee, John. *The Pine Barrens*. Special edition with photographs by Bill Curtsinger. New York: Farrar Straus Giroux, 1981.

Meadows, Donella. "Nothing Is so Powerful as an Exponential Whose Time Has Come." The Donella Meadows Project, Academy for Systems Change, June 16, 1989,

Donella Meadows Archives. http://donellameadows.org/archives/nothing-is-so
-powerful-as-an-exponential-whose-time-has-come/.

Medina, José. *The Epistemology of Resistance: Gender and Racial Oppression, Epistemic Injustice, and the Social Imagination.* Studies in Feminist Philosophy. New York: Oxford University Press, 2013.

"Members of the Coronavirus Task Force Hold a Press Briefing." Trump White House Archived, YouTube video, 43:10, March 10, 2020. https://www.youtube.com/watch?v=RLcyNzloAwM.

Menely, Tobias. "Anthropocene Air." *Minnesota Review* 2014, no. 83 (November 1, 2014): 93–101.

Middel, Ariane, Nalini Chhetri, and Raymond Quay. "Urban Forestry and Cool Roofs: Assessment of Heat Mitigation Strategies in Phoenix Residential Neighborhoods." *Urban Forestry & Urban Greening* 14, no. 1 (January 1, 2015): 178–86.

Miller, Carolyn. "The Aristotelian *Topos*: Hunting for Novelty." In *Rereading Aristotle's* Rhetoric, edited by Allan G. Gross and Arthur E. Walzer, 130–48. Carbondale: Southern Illinois University Press, 2000.

Miller, Carolyn R. "Aristotle's 'Special Topics,' in Rhetorical Practice and Pedagogy." *Rhetoric Society Quarterly* 17, no. 1 (January 1, 1987): 61–70.

Millman, Jennifer. "Midtown Lawyer Positive for Coronavirus Is NY's First Case of Person-to-Person Spread." *Channel 4 NBC New York News*, March 3, 2020. https://www.nbcnewyork.com/news/local/westchester-county-man-tests-positive-for-coronavirus-in-nys-1st-possible-community-spread-case-gov-cuomo/2310134/.

Milman, Oliver, and David Smith. "'Listen to the Scientists': Greta Thunberg Urges Congress to Take Action." *Guardian*, September 19, 2020. https://www.theguardian.com/us-news/2019/sep/18/greta-thunberg-testimony-congress-climate-change-action.

Mitchell, Daniel M., Scott M. Osprey, Lesley J. Gray, Neal Butchart, Steven C. Hardiman, Andrew J. Charlton-Perez, and Peter Watson. "The Effect of Climate Change on the Variability of the Northern Hemisphere Stratospheric Polar Vortex." *Journal of the Atmospheric Sciences* 69, no. 8 (August 1, 2012): 2608–18.

Moffitt, Evan. "Maya Lin Builds a Final Monument to the Climate." *Frieze*, April 28, 2021. https://www.frieze.com/article/maya-lin-ghost-forest-2021.

Monbiot, George. *Heat: How to Stop the Planet from Burning.* Cambridge: South End Press, 2007.

Mooney, Chris. "The Hockey Stick: The Most Controversial Chart in Science, Explained." *Atlantic*, May 10, 2013. https://www.theatlantic.com/technology/archive/2013/05/the-hockey-stick-the-most-controversial-chart-in-science-explained/275753/.

Moriarty, Devon, Paula Núñez De Villavicencio, Lillian A. Black, Monica Bustos, Helen Cai, Brad Mehlenbacher, and Ashley Rose Mehlenbacher. "Durable Research, Portable Findings: Rhetorical Methods in Case Study Research." *Technical Communication Quarterly* 28, no. 2 (April 3, 2019): 124–36.

Mortimer-Sandilands, Catriona. "Melancholy Natures, Queer Ecologies." In *Queer Ecologies: Sex, Nature, Politics, Desire,* edited by Catriona Mortimer-Sandilands and Bruce Erickson, 331–58. Bloomington: Indiana University Press, 2010.

Morton, Timothy. *Being Ecological.* New York: Penguin Books Limited, 2018.

Morton, Timothy. *Dark Ecology: For a Logic of Future Coexistence.* New York: Columbia University Press, 2016.

Morton, Timothy. *Hyperobjects: Philosophy and Ecology after the End of the World*. Minneapolis: University of Minnesota Press, 2013.

MoveOn.org. "Commit to #FlattenTheCurve." MoveOn.org, accessed September 17, 2020. https://act.moveon.org/survey/flatten-the-curve.

Murphy, John M. "Presence, Analogy, and Earth in the Balance." *Argumentation and Advocacy* 31, no. 1 (June 1, 1994): 1–16.

Murphy, Patrick D. "Speaking for the Youth, Speaking for the Planet: Greta Thunberg and the Representational Politics of Eco-Celebrity." *Popular Communication* 19, no. 3 (July 3, 2021): 193–206.

NASA. "Carbon Dioxide, Latest Measurement, June 2020." Global Climate Change: Vital Signs of the Planet, accessed September 23, 2020. https://climate.nasa.gov/vital-signs/carbon-dioxide/.

National Oceanographic and Atmospheric Administration. "The Keeling Curve and the NOAA Climate Model." Global Monitoring Laboratory, accessed May 12, 2022. https://gml.noaa.gov/ccgg/trends/.

National Oceanographic and Atmospheric Administration. "What Is a Ghost Forest?" National Ocean Service. Accessed April 20, 2021. https://oceanservice.noaa.gov/facts/ghost-forest.html.

National Snow and Ice Data Center. "Dead Ice." *Cryosphere Glossary*, accessed November 18, 2019. https://nsidc.org/cryosphere/glossary/term/dead-ice.

National Snow and Ice Data Center. "Thermodynamics: Albedo." All About Sea Ice, last updated April 3, 2020. https://nsidc.org/cryosphere/seaice/processes/albedo.html.

Nealon, Jeffrey T. *Plant Theory: Biopower and Vegetable Life*. Stanford, CA: Stanford University Press, 2016.

Netburn, Deborah. "From COVID to Curbside, 2020 Changed Our Vocabulary Too." *Los Angeles Times*, December 28, 2020. https://www.latimes.com/science/story/2020-12-28/from-covid-to-curbside-2020-changed-our-vocabulary-too.

Nichols, Ann Eljenholm. "The Suasive Subjunctive: Alive and Well in the Upper Midwest." *American Speech* 62, no. 2 (1987): 140–53.

Nicotra, Jodie, and Judith Totman Parrish. "Rushing the Cure: Temporal Rhetorics in Global Warming Discourse." *JAC* 30, no. 1/2 (2010): 215–37.

Nisbet, Matthew C. "Communicating Climate Change: Why Frames Matter for Public Engagement." *Environment: Science and Policy for Sustainable Development* 51, no. 2 (March 1, 2009): 12–23.

Nisbet, Matthew C., and Chris Mooney. "Framing Science." *Science* 316, no. 5821 (April 6, 2007): 56–56.

Nisbet, Matthew C., and Dietram A. Scheufele. "What's Next for Science Communication? Promising Directions and Lingering Distractions." *American Journal of Botany* 96, no. 10 (2009): 1767–78.

Nixon, Rob. *Slow Violence and the Environmentalism of the Poor*. Cambridge, MA: Harvard University Press, 2011.

Nussbaum, Martha C. *The Monarchy of Fear: A Philosopher Looks at Our Political Crisis*. New York: Simon & Schuster, 2018.

Oakley, Todd. *From Attention to Meaning: Explorations in Semiotics, Linguistics, and Rhetoric*. European Semiotics. Bern, Switzerland: Peter Lang, 2009.

Obama, Barack. "Remarks by the President at Exit Glacier, Kenai Fjords National Park, AK." White House Office of the Press Secretary, September 1, 2015. https://

obamawhitehouse.archives.gov/the-press-office/2015/09/01/remarks-president
-exit-glacier-kenai-fjords-national-park-ak.

Obama, Barack. "Just 16, Greta Thunberg Is Already One of Our Planet's Greatest Advocates." Twitter post, September 17, 2019. 4:24 PM. https://twitter.com/Barack Obama/status/1174056583610949632.

Obrist, Hans Ulrich. "The Future Is Invented with Fragments from the Past." In *The Oldest Living Things in the World*, xiv–xvii. Chicago: University of Chicago Press, 2014.

O'Dowd, Peter. "Researchers Memorialize First Major Icelandic Glacier Lost to Climate Change." *Here and Now*, WBUR, August 16, 2019. https://www.wbur.org /hereandnow/2019/08/16/dead-icelandic-glacier-lost-to-climate-change.

O'Gorman, Ned. "Aristotle's *Phantasia* in the *Rhetoric*: *Lexis*, Appearance, and the Epideictic Function of Discourse." *Philosophy and Rhetoric* 38, no. 1 (2005): 16–40.

O'Hagan, Andrew. "Keepsakes." *New York Times Magazine*, July 12, 2020, 30–31.

Oliver, Kelly. *Witnessing: Beyond Recognition*. Minneapolis: University of Minnesota Press, 2001.

Olson, Christa J. "American Magnitude: Frederic Church, Hiram Bingham, and Hemispheric Vision." *Rhetoric Society Quarterly* 48, no. 4 (August 8, 2018): 380–404.

Olson, Christa. *American Magnitude: Hemispheric Vision and Public Feeling in the United States*. Columbus: Ohio State University Press, 2021.

Olson, Kathryn M. "Rhetorical Leadership and Transferable Lessons for Successful Social Advocacy in Al Gore's *An Inconvenient Truth*." *Argumentation and Advocacy* 44, no. 2 (September 1, 2007): 90–109.

Oltof, Jelte. "Projecting a Future Present: Greta Thunberg's Use of Presence at the United Nations Climate Action Summit 2019." *Res Rhetorica* 8, no. 1 (2021): 67–82.

Onder, Graziano, Giovanni Rezza, and Silvio Brusaferro. "Case-Fatality Rate and Characteristics of Patients Dying in Relation to Covid-19 in Italy." *JAMA*, March 23, 2020.

O'Neill, Saffron J. "Image Matters: Climate Change Imagery in US, UK and Australian Newspapers." *Geoforum* 49 (October 1, 2013): 10–19.

O'Neill, Saffron J., and Nicholas Smith. "Climate Change and Visual Imagery." *WIREs Climate Change* 5, no. 1 (January 1, 2014): 73–87.

O'Neill, Saffron, and Sophie Nicholson-Cole. "'Fear Won't Do It': Promoting Positive Engagement with Climate Change through Visual and Iconic Representations." *Science Communication* 30, no. 3 (January 7, 2009): 355–79.

Onís, Catalina M. de. *Energy Islands: Metaphors of Power, Extractivism, and Justice in Puerto Rico*. Oakland: University of California Press, 2021.

Our Children's Trust. "Legal Proceedings in All 50 States." State Legal Actions, accessed January 3, 2021. https://www.ourchildrenstrust.org/other-proceedings-in-all -50-states.

Parry, Wayne. "Seas Rise, Trees Die: Climate Change Before Your Eyes." *Concord Monitor*, August 1, 2017. https://www.concordmonitor.com/Seas-rise-trees-die-Climate -change-before-your-eyes-11617944.

Patterson, Angelica E. "Warming Climates, Changing Forests." *Consilience* 9 (February 2013): 160–71.

Perelman, Chaim. *Justice, Law, and Argument: Essays on Moral and Legal Reasoning*. Synthese Library. Dordrecht, Holland: D. Reidel Publishing Company, 1980.

Perelman, Chaim. *The New Rhetoric and the Humanities: Essays on Rhetoric and Its Applications*. Dordrecht, Holland: D. Reidel Publishing Company, 1979.

Perelman, Chaim, and Lucie Olbrechts-Tyteca. *The New Rhetoric: A Treatise on Argumentation*. Translated by John Wilkinson and Purcell Weaver. Notre Dame, IN: University of Notre Dame Press, 1969.

Pernot, Laurent. *Epideictic Rhetoric: Questioning the Stakes of Ancient Praise*. Austin: University of Texas Press, 2015.

Peters, Jason. "Public Art as Social Infrastructure: Methods and Materials for Social Action at Environmentally Contaminated Sites." *Reflections* 19, no. 2 (Fall/Winter, 2020): 106–29.

Peters, John Durham. "33 + 1 Vignettes on the History of Scalar Inversion." *ELH* 86, no. 2 (2019): 305–31.

Peters, John Durham. *Courting the Abyss: Free Speech and the Liberal Tradition*. Chicago: University of Chicago Press, 2005.

Peters, John Durham. *The Marvelous Clouds: Toward a Philosophy of Elemental Media*. Chicago: University of Chicago Press, 2015.

Peters, John Durham. "Witnessing." *Media, Culture & Society* 23, no. 6 (November 1, 2001): 707–23.

Peterson, Tarla Rai. "The Rhetorical Construction of Institutional Authority in a Senate Subcommittee Hearing on Wilderness Legislation." *Western Journal of Speech Communication* 52, no. 4 (December 30, 1988): 259–76.

Pezzullo, Phaedra C. "Environment." In *The Oxford Encyclopedia of Communication and Critical Cultural Studies*, 573–89. New York: Oxford University Press, 2019.

Pezzullo, Phaedra C. *Toxic Tourism: Rhetorics of Pollution, Travel, and Environmental Justice*. Tuscaloosa: University of Alabama Press, 2009.

Pezzullo, Phaedra C., and Catalina M. de Onís. "Rethinking Rhetorical Field Methods on a Precarious Planet." *Communication Monographs* 85, no. 1 (January 2, 2018): 103–22.

Pezzullo, Phaedra C., and Robert Cox. *Environmental Communication and the Public Sphere*. 3rd edition. Los Angeles: Sage Publications, 2013.

Pfau, Michael William. "Who's Afraid of Fear Appeals? Contingency, Courage, and Deliberation in Rhetorical Theory and Practice." *Philosophy & Rhetoric* 40, no. 2 (2007): 216–37.

Phillips, Kendall R. "The Failure of Memory: Reflections on Rhetoric and Public Remembrance." *Western Journal of Communication* 74, no. 2 (April 1, 2010): 208–23.

Phillips, Kendall R., and G. Mitchell Reyes. "Introduction: Surveying Global Memoryscapes: The Shifting Terrain of Public Memory Studies." In *Global Memoryscapes: Contesting Remembrance in a Transnational Age*, edited by Kendall R. Phillips and G. Mitchell Reyes, 1–26. Tuscaloosa: University of Alabama Press, 2011.

Picard, Caroline. "An Opening to Imagine the Present: A Conversation with Cymene Howe and Anand Pandian." Blog post, August 24, 2016. http://badatsports.com /2016/an-opening-to-imagine-the-present-a-conversation-with-cymene-howe-and -anand-pandian/.

Poggioli, Sylvia. "Lockdown, Safety Measures Bear Success in Italy." *All Things Considered*, National Public Radio, June 27, 2020. https://www.npr.org/2020/06/27 /884307558/lockdown-safety-measures-bear-success-in-italy.

Pogrebin, Robin. "Maya Lin's 'Ghost Forest' Is a Climate Warning." *New York Times*, November 14, 2019, sec. C.

Pongsiri, Montira J., Joe Roman, Vanessa O. Ezenwa, Tony L. Goldberg, Hillel S. Koren, Stephen C. Newbold, Richard S. Ostfeld, Subhrendu K. Pattanayak, and Daniel J. Salkeld. "Biodiversity Loss Affects Global Disease Ecology." *BioScience* 59, no. 11 (December 1, 2009): 945–54.

Poole, Megan. "Orientation: Seeing and Sensing Rhetorically." *Western Journal of Communication* 84, no. 5 (2020): 604–22.

Poole, Megan. "Technical Beauty: Rhetorics and Aesthetics of Science." PhD diss., The Pennsylvania State University, 2020.

Poole, Megan. "Witnessing the Open Semiosis: A Method for Rhetorical Listening Beyond the Human." *Rhetoric Society Quarterly*, forthcoming, 2022.

Poulakos, John. "Toward a Sophistic Definition of Rhetoric." *Philosophy & Rhetoric* 16, no. 1 (1983): 35–48.

Powers, Richard. *The Overstory*. New York: W. W. Norton, 2018.

Pozo, Carlos, Ángel Galán-Martín, David M. Reiner, Niall Mac Dowell, and Gonzalo Guillén-Gosálbez. "Equity in Allocating Carbon Dioxide Removal Quotas." *Nature Climate Change* 10, no. 7 (July 1, 2020): 640–46.

Prasch, Allison. "Toward a Rhetorical Theory of Deixis." *Quarterly Journal of Speech* 102, no. 2 (2016): 166–93.

Predmore, S. Andrew, Josh McDaniel, and John S. Kush. "Presettlement Forests and Fire in Southern Alabama." *Canadian Journal of Forest Research* 37, no. 9 (September 1, 2007): 1723–36.

Prior, Paul. *Writing/Disciplinarity: A Sociohistoric Account of Literate Activity in the Academy*. New York: Lawrence Erlbaum Associates, 1998.

Prior, Paul, and Jody Shipka. "Chronotopic Lamination: Tracing the Contours of Literate Activity." In *Writing Selves, Writing Societies: Research from Activity Perspectives*, 180–238. Fort Collins, CO: The WAC Clearinghouse, 2003.

Quiñones Valdivia, Fernando Ismael. "Text as a Nowing: Towards an Understanding of Time in Rhetoric." *Quarterly Journal of Speech* 107, no. 2 (2021): 141–59.

Radway, Janice. "Foreword." In *Ghostly Matters: Haunting and the Sociological Imagination*, vii–xiii. Minneapolis: University of Minnesota Press, 2009.

Rai, Candice. *Democracy's Lot: Rhetoric, Publics, and the Places of Invention*. Tuscaloosa: University of Alabama Press, 2016.

Ramanathan, V., R. Aines, M. Auffhammer, M. Barth, J. Cole, and F. Forman. *Bending the Curve: Climate Change Solutions*. Edited by V. Ramanathan, Adam Millard-Ball, Michelle Niemann, and Scott Friese. Oakland: University of California Press, 2019. https://escholarship.org/uc/item/6kr8p5rq.

Rapaport, Brooke Kamin. "Maya Lin's 'Ghost Forest.'" Audio tour, May 10, 2021. https://madisonsquarepark.org/art/exhibitions/maya-lin-ghost-forest/.

Raskin, Paul D. "Bending the Curve: Toward Global Sustainability." *Development* 43, no. 4 (December 1, 2000): 67–74.

Reilly, Kate H., and Jan F. Adamowski. "Spatial and Temporal Scale Framing of a Decision on the Future of the Mactaquac Dam in New Brunswick, Canada." *Ecology and Society* 22, no. 3 (2017).

Reyes, G. Mitchell. *The Evolution of Mathematics: A Rhetorical Approach*. University Park: Pennsylvania State University Press, 2022.

Reyes, G. Mitchell. "The Horizons of Judgment in Mathematical Discourse: Copulas, Economics, and Subprime Mortgages." In *Arguing with Numbers: The Intersections of Rhetoric and Mathematics*, edited by James Wynn and G. Mitchell Reyes, 82–121. University Park: Pennsylvania State University Press, 2021.

Reyes, G. Mitchell. "Stranger Relations: The Case for Rebuilding Commonplaces between Rhetoric and Mathematics." *Rhetoric Society Quarterly* 44, no. 5 (October 20, 2014): 470–91.

Rice, Doyle. "Greta Thunberg Is Sailing Across the Atlantic. Here's What She Accom

plished While in the US." *USA Today,* November 13, 2019. https://www.usatoday
.com/story/news/nation/2019/11/13/climate-activist-greta-thunberg-sails-back
-europe-la-vagabonde/4178195002/.

Rice, Jenny. *Awful Archives: Conspiracy Theory, Rhetoric, and Acts of Evidence.* Columbus: Ohio State University Press, 2020.

Rice, Jenny. *Distant Publics: Development Rhetoric and the Subject of Crisis.* Pittsburgh,
PA: University of Pittsburgh Press, 2012.

Rice, Jenny. "The Rhetorical Aesthetics of More: On Archival Magnitude." *Philosophy
& Rhetoric* 50, no. 1 (2017): 26–49.

Rice University News and Media. "Lost Glacier to Be Honored with Memorial Monument." News release, July 18, 2019.

Rickert, Thomas. *Ambient Rhetoric: The Attunements of Rhetorical Being.* Pittsburgh,
PA: University of Pittsburgh Press, 2013.

Roald, Veborn, and Linda Sangolt. *Deliberation, Rhetoric, and Emotion in the Discourse
on Climate Change in the European Parliament.* Delft, Netherlands: Eburon, 2012.

Roberts, Siobhan. "Flattening the Coronavirus Curve." *New York Times,* March 27,
2020. https://www.nytimes.com/article/flatten-curve-coronavirus.html.

Root, Tik. "Ghost Forests Are Visceral Examples of the Advance of Climate Change."
Time, October 7, 2019. https://time.com/5694648/ghost-forests-climate-change/.

Ross, Christine. "The Aesthetics of Coexistence as Ongoing." In *Aesthetic Temporalities Today: Present, Presentness, Re-Presentation,* edited by Gabriele Genge, Ludger
Schwarte, and Angela Stercken, 169–82. Bielefeld, Germany: transcript Verlag, 2020.

Rosteck, Thomas, and Thomas S. Frentz. "Myth and Multiple Readings in Environmental Rhetoric: The Case of *An Inconvenient Truth.*" *Quarterly Journal of Speech* 95,
no. 1 (February 1, 2009): 1–19.

Rothberg, Michael. "After Apartheid, Beyond Filiation: Witnessing and the Work of
Justice." *Law & Literature* 21, no. 2 (June 1, 2009): 275–90.

Rothberg, Michael. "Between Auschwitz and Algeria: Multidirectional Memory and
the Counterpublic Witness." *Critical Inquiry* 33, no. 1 (September 1, 2006): 158–84.

Rothfield, Michael, Somini Sengupta, Joseph Goldstein, and Brian M. Rosenthal. "13
Deaths in a Day: An 'Apocalyptic' Coronavirus Surge at an NYC Hospital." *New
York Times,* March 25, 2020. https://www.nytimes.com/2020/03/25/nyregion/nyc
-coronavirus-hospitals.html.

"Rules of the One Hundred Sixteenth Congress," H.Res. 6, January 9, 2019, 43–44.

Ryalls, Emily D., and Sharon R. Mazzarella. "'Famous, Beloved, Reviled, Respected,
Feared, Celebrated': Media Construction of Greta Thunberg." *Communication, Culture and Critique* 14, no. 3 (2021): 438–53.

Sæþórsdóttir, Anna Dóra, and Jarkko Saarinen. "Challenges Due to Changing Ideas of
Natural Resources: Tourism and Power Plant Development in the Icelandic Wilderness." *Polar Record* 52, no. 1 (2016): 82–91.

Schieder, Nathalie W., and Matthew L. Kirwan. "Sea-Level Driven Acceleration in
Coastal Forest Retreat." *Geology* 47, no. 12 (October 2, 2019): 1151–55.

Schneider, Birgit. "Image Politics: Picturing Uncertainty. The Role of Images in Cli-
matology and Climate Policy." In *Climate Change and Policy: The Calculability of
Climate Change and the Challenge of Uncertainty,* edited by Gabriele Gramelsberger
and Johann Feichter, 191–209. Berlin: Springer, 2011.

Schneider, Birgit, and Thomas Nocke. "Image Politics of Climate Change: Introduction." In *Image Politics of Climate Change: Visualizations, Imaginations, Documenta-
tions,* 9–25. Image. Bielefeld, Germany: transcript Verlag, 2014.

Schneider, Birgit, and Lynda Walsh. "The Politics of Zoom: Problems with Downscaling Climate Visualizations." *Geo: Geography and Environment* 6, no. 1 (2019): 1–11.

Schroeder, James G., Michael A. Taras, and United States Forest Service. *Atlantic White-Cedar: (Chamaecyparis thyoides (L.) B.S.P.).* [Washington, DC]: Forest Service, United States Department of Agriculture, 1985.

Schroth, Olaf, Jeannette Angel, Stephen Sheppard, and Aleksandra Dulic. "Visual Climate Change Communication: From Iconography to Locally Framed 3D Visualization." *Environmental Communication* 8, no. 4 (October 2, 2014): 413–32.

Schulgasser, K., and A. Witztum. "The Mechanism of Spiral Grain Formation in Trees." *Wood Science and Technology* 41, no. 2 (September 12, 2006): 133–56.

Schryer, Catherine F. "Genre Time/Space: Chronotopic Strategies in the Experimental Article." *JAC* 19, no. 1 (1999): 81–89.

Schutten, Julie Kalil. "Environmental Sustainability: Witnessing, Embodiment, and the Grotesque." *JAC* 31, no. 1/2 (2011): 338–49.

Scott, Sir Walter. *The Journal of Sir Walter Scott.* Edited by W. E. K. Anderson. Oxford: Oxford University Press, 1972.

Seeger, Matthew W., and Timothy L. Sellnow. *Narratives of Crisis: Telling Stories of Ruin and Renewal.* High Reliability and Crisis Management. Stanford, CA: Stanford University Press, 2016.

"Seize the Moment." Editors. *The Economist*, May 23, 2020.

Seligman, Adam B. "Ritual and Sincerity: Certitude and the Other." *Philosophy & Social Criticism* 36, no. 1 (December 30, 2009): 9–39.

Sengupta, Somini. "Greta Thunberg, on Tour in America, Offers an Unvarnished View." *New York Times*, September 18, 2019. https://www.nytimes.com/2019/09/18/climate/greta-thunberg.html.

Sengupta, Somini. "Greta Thunberg Sets Sail, Again, After Climate Talks Relocate." *New York Times*, November 12, 2019. https://www.nytimes.com/2019/11/12/climate/greta-thunberg-return-europe.html.

Shapiro, Mark. "Michael Mann on Climate Change and Speaking Out." Explore the Space podcast, episode 152, September 23, 2019. https://www.explorethespaceshow.com/podcasting/michael-mann-on-climate-change-speaking-out/.

Sharma, Sarah. *In the Meantime: Temporality and Cultural Politics.* Durham, NC: Duke University Press, 2014.

Shaw, Christopher. "The Two Degrees Celsius Limit." In *Oxford Research Encyclopedia of Climate Science.* Oxford: Oxford University Press, 2017. https://oxfordre.com/climatescience/view/10.1093/acrefore/9780190228620.001.0001/acrefore-9780190228620-e-15.

Sheard, Cynthia L. "The Public Value of Epideictic Rhetoric." *College English* 58, no. 7 (1996): 765–94.

Sheikh, Shela. "The Future of the Witness: Nature, Race and More-than-Human Environmental Publics." *Kronos* 44, no. 1 (2018): 145–62.

Sheppard, Stephen R. J. "Landscape Visualisation and Climate Change: The Potential for Influencing Perceptions and Behaviour." *Environmental Science & Policy* 8, no. 6 (2005): 637–54.

Shilling, Dan. "Introduction: The Soul of Sustainability." In *Traditional Ecological Knowledge: Learning from Indigenous Practices for Environmental Sustainability*, edited by Melissa K. Nelson and Dan Shilling, 1–14. New Directions in Sustainability and Society. Cambridge: Cambridge University Press, 2018.

Shumskiy, P. A. "Density of Glacier Ice." *Journal of Glaciology* 3, no. 27 (1960): 568–73.

Simard, Suzanne. *Finding the Mother Tree: Discovering the Wisdom of the Forest*. New York: Knopf Doubleday Publishing Group, 2021.

Simard, Suzanne. "How Trees Talk to Each Other." TEDSummit, June 2016. https://www.ted.com/talks/suzanne_simard_how_trees_talk_to_each_other?language=en.

Small, Zachary. "Maya Lin's Dismantled 'Ghost Forest' to Be Reborn as Boats." *New York Times*, November 24, 2021. https://www.nytimes.com/2021/11/24/arts/design/maya-lin-rocking-the-boat.html.

Smith, John E. "Time and Qualitative Time." In *Rhetoric and "Kairos": Essays in History, Theory, and Praxis*, edited by Phillip Sipiora and James S. Baumlin, 46–57. Albany: State University of New York Press, 2002.

Smith, John E. "Time, Times, and the 'Right Time': Chronos and Kairos." *The Monist* 53, no. 1 (January 1, 1969): 1–13.

Song, Young Imm Kang. "A New Vision for Public Art and Functional Landscape Design." *International Journal of Art & Design Education* 33, no. 2 (June 1, 2014): 242–57.

St. Amant, Kirk, and S. Scott Graham. "Research that Resonates: A Perspective on Durable and Portable Approaches to Scholarship in Technical Communication and Rhetoric of Science." *Technical Communication Quarterly* 28, no. 2 (April 3, 2019): 99–111.

Steffen, Will, Paul J. Crutzen, and John R. McNeill. "The Anthropocene: Are Humans Now Overwhelming the Great Forces of Nature?" *Ambio* 36, no. 8 (2007): 614–21.

Stein, Theo. "Rise of Carbon Dioxide Unabated: Seasonal Peak Reaches 417 Parts per Million at Mauna Loa Observatory." *NOAA Research News*, June 4, 2020. https://research.noaa.gov/article/ArtMID/587/ArticleID/2636/Rise-of-carbon-dioxide-unabated#:~:text=Seasonal%20peak%20reaches%20417%20parts%20per%20million%20at%20Mauna%20Loa%20observatory&text=Scripps%20scientists%20reported%20an%20May,atmosphere%20in%20several%20million%20years.

Stephens, Andrew J., Derek Clarke, and Robert J. Nicholls. "Trends in Reported Flooding in the UK: 1884–2013." *Hydrological Sciences Journal* 61, no. 1 (January 2, 2016): 53–55.

Sterman, John D. "Communicating Climate Change Risks in a Skeptical World." *Climatic Change* 108, no. 4 (August 18, 2011): 811–26.

Stewart, Kathleen. "Weak Theory in an Unfinished World." *Journal of Folklore Research* 45, no. 1 (2008): 71–82.

Strutt, Jacob George. *Sylva Britannica; or, Portraits of Forest Trees, Distinguished for Their Antiquity, Magnitude, or Beauty Drawn from Nature*. Folio Edition. London: published by the author, 1830. https://archive.org/details/sylvabritannica00strurich/page/xiv/mode/2up?view=theater.

Sturken, Marita. "The Wall, the Screen, and the Image: The Vietnam Veterans Memorial." *Representations* 35, special issue: Monumental Histories (1991): 118–42.

Sussman, Rachel. "Oldest Living Things in the World." *Rachel Sussman / Projects* (artist site), n.d. http://www.rachelsussman.com/oltw.

Sussman, Rachel. *The Oldest Living Things in the World*. Chicago: University of Chicago Press, 2014.

Swant, Marty. "Reddit Cofounder Alexis Ohanian Bought a Billboard in Times Square to Inform People about COVID-19." *Forbes*, March 14, 2020. https://www.forbes.com/sites/martyswant/2020/03/14/reddit-co-founder-alexis-ohanian-bought-a-billboard-in-time-square-to-inform-people-about-covid-19/#2ae1b2d73a13.

Szerszynski, Bronislaw. "Ecological Rites: Ritual Action in Environmental Protest Events." *Theory, Culture & Society* 19, no. 3 (June 1, 2002): 51–69.

Taylor, Benton N., Angelica E. Patterson, Moyosore Ajayi, Rachel Arkebauer, Karen Bao, Natalie Bray, Robert M. Elliott, et al. "Growth and Physiology of a Dominant Understory Shrub, *Hamamelis virginiana*, Following Canopy Disturbance in a Temperate Hardwood Forest." *Canadian Journal of Forest Research* 47, no. 2 (February 1, 2017): 193–202.

Taylor, Derrick Bryson. "Trump Mocks Greta Thunberg on Twitter, and She Jabs Back." *New York Times*, December 12, 2019. https://www.nytimes.com/2019/12/12/us/politics/greta-thunberg-trump.html.

Thunberg, Greta. "Transcript: Greta Thunberg's Speech at the UN Climate Action Summit." National Public Radio, September 23, 2019. https://www.npr.org/2019/09/23/763452863/transcript-greta-thunbergs-speech-at-the-u-n-climate-action-summit#:~:text=People%20are%20suffering.,tales%20of%20eternal%20economic%20growth.

Thunberg, Greta, and George Monbiot. "Greta Thunberg and George Monbiot Make Short Film on the Climate Crisis." YouTube video, 3:40, September 19, 2019. https://www.youtube.com/watch?v=-Q0xUXo2zEY&t=1s.

Tirmenstein, D. A. "*Chamaecyparis thyoides* in Fire Effects Information System." Rocky Mountain Research Station, US Department of Agriculture, Forest Service, 1991. https://www.fs.fed.us/database/feis/plants/tree/chathy/all.html.

Tollefson, Jeff, and Emiliano Rodríguez Mega. "Geoscientist Faces Criminal Charges Over Glacier Survey." *Nature* 552 (December 14, 2017): 159–60.

Torcello, Lawrence, and Michael E. Mann. "Seeing the COVID-19 Crisis Is Like Watching a Time Lapse of Climate Change. Will the Right Lessons Be Learned?" *Newsweek*, April 1, 2020. https://www.newsweek.com/fake-news-climate-change-coronavirus-time-lapse-1495603.

Tronto, Joan. *Caring Democracy: Markets, Equality, and Justice.* New York: New York University Press, 2013.

Tronto, Joan. *Moral Boundaries: A Political Argument for an Ethic of Care.* New York: Routledge, 1993.

Trouet, Valerie. *Tree Story: The History of the World Written in Rings.* Baltimore, MD: Johns Hopkins University Press, 2020.

Tsing, Anna Lowenhaupt. *The Mushroom at the End of the World: On the Possibility of Life in Capitalist Ruins.* Princeton, NJ: Princeton University Press, 2015.

Tuck, Eve, and C. Ree. "A Glossary of Haunting." In *Handbook of Autoethnography*, 639–58. New York: Routledge, 2013.

Tufte, Edward R. *The Visual Display of Quantitative Information.* Cheshire, CT: Graphics Press, 1983.

Tufte, Edward R. *Visual Explanations: Images and Quantities, Evidence and Narrative.* Cheshire, CT: Graphics Press, 1997.

Tversky, Amos, and Daniel Kahneman. "Availability: A Heuristic for Judging Frequency and Probability." *Cognitive Psychology* 5, no. 2 (September 1, 1973): 207–32.

United Nations. "Emissions Gap Report 2019." Nairobi: UN Environmental Program, 2019.

United States Department of Agriculture and Natural Resources Conservation Service. "New Jersey Fact Sheet: Atlantic White Cedar." Accessed July 18, 2022. https://njaudubon.org/wp-content/uploads/2019/09/AWC_and_Forest_Management_Fact_Sheet_NJAS.pdf.

US Surgeon General. "A Little #MondayMotivation for Those Frustrated with Mitigation." Twitter post, @Surgeon_General, April 20, 2020. https://twitter.com/Surgeon_General/status/1252206970028908547.

Van Lieshout, Maartje, Art Dewulf, Noelle Aarts, and Catrien Termeer. "The Power to Frame the Scale? Analysing Scalar Politics over, in and of a Deliberative Governance Process." *Journal of Environmental Policy & Planning* 19, no. 5 (September 3, 2017): 550–73.

Van Winkle, Kevin. "Above All Made by Themselves: The Visual Rhetoric of W. E. B. Du Bois's Data Visualizations." *Technical Communication Quarterly* 1 (March 21, 2021): 1–16.

Verhulsdonck, Gustav, and Vishal Shah. "Lean Data Visualization: Considering Actionable Metrics for Technical Communication." *Journal of Business and Technical Communication* 35, no. 1 (January 1, 2021): 57–64.

Vineyard Gazette. "Rare Footage of Heath Hen on Martha's Vineyard." YouTube video, 2:59, July 17, 2014. https://www.youtube.com/watch?v=FOFphkQnRqE.

Vivian, Bradford. *Commonplace Witnessing: Rhetorical Invention, Historical Remembrance, and Public Culture.* New York: Oxford University Press, 2017.

Vivian, Bradford. "Epideictic Forgetting in Booker T. Washington's Cotton States Exposition Address." *Philosophy & Rhetoric* 45, no. 2 (2012): 189–212.

Vivian, Bradford. "Neoliberal Epideictic: Rhetorical Form and Commemorative Politics on September 11, 2002." *Quarterly Journal of Speech* 92, no. 1 (February 2006): 1–26.

Vivian, Bradford. "Witnessing Time: Rhetorical Form, Public Culture, and Popular Historical Education." *Rhetoric Society Quarterly* 44, no. 3 (May 27, 2014): 204–19.

Von Bergen, Megan, and Bethany Mannon. "Talking Climate Faith: Katharine Hayhoe and Christian Rhetoric(s) of Climate Change." *Enculturation: A Journal of Rhetoric, Writing, and Culture*, November 10, 2020. http://enculturation.net/Talking%20Climate%20Faith.

Wagenaar, William A., and Sabato D. Sagaria. "Misperception of Exponential Growth." *Perception & Psychophysics* 18, no. 6 (November 1, 1975): 416–22.

Waldman, John. "With Temperatures Rising, Here Comes 'Global Weirding.'" *Yale E360*, March 19, 2009. https://e360.yale.edu/features/with_temperatures_rising_here_comes_global_weirding.

Walker, Jeffrey. *Rhetoric and Poetics in Antiquity.* New York: Oxford University Press, 2000.

Walker, Kenny, and Lynda Walsh. "'No One Yet Knows What the Ultimate Consequences May Be': How Rachel Carson Transformed Scientific Uncertainty into a Site for Public Participation in *Silent Spring*." *Journal of Business and Technical Communication* 26, no. 1 (December 7, 2011): 3–34.

Walsh, Lynda. *Scientists as Prophets: A Rhetorical Genealogy.* New York: Oxford University Press, 2013.

Walsh, Lynda. "The Visual Rhetoric of Climate Change." *Wiley Interdisciplinary Reviews: Climate Change* 6, no. 4 (2015): 361–68.

Walsh, Lynda, and Kenneth C. Walker. "Perspectives on Uncertainty for Technical Communication Scholars." *Technical Communication Quarterly* 25, no. 2 (April 2, 2016): 71–86.

Welling, Johannes T. "Tourism, Landscapes, and Climate Change in Iceland." Icelandic Tourism Research Centre, November 2013.

Weston, Kath. *Animate Planet: Making Visceral Sense of Living in a High-Tech Ecologically Damaged World.* Durham, NC: Duke University Press, 2017.

Whale, Helen, and Franklin Ginn. "In the Absence of Sparrows." In *Mourning Nature: Hope at the Heart of Ecological Loss and Grief*, edited by Ashlee Cunsolo and Karen Landman, 92–116. Montreal: McGill-Queen's University Press, 2017.

"What Can We Learn from the Coronavirus Outbreak that Could Help Us Fight Climate Change?" Sustainability for All. https://www.activesustainability.com/climate -change/learn-coronavirus-outbreak-fight-climate-change/?_adin=02021864894.

Whyte, Kyle. "Against Crisis Epistemology." In *Routledge Handbook of Critical Indigenous Studies*, edited by Aileen Moreton-Robinson, Linda Tuhiwai-Smith, Chris Andersen, and Steve Larkin, 52–64. Routledge International Handbooks. New York: Taylor & Francis, 2022.

Whyte, Kyle Powys. "On the Role of Traditional Ecological Knowledge as a Collaborative Concept: A Philosophical Study." *Ecological Processes* 2, no. 1 (April 5, 2013): 7.

Whyte, Kyle. "Time as Kinship." In *The Cambridge Companion to the Environmental Humanities*, edited by Jeffrey Jerome Cohen and Stephanie Foote, 39–55. Cambridge: Cambridge University Press, 2021.

Whyte, Kyle, Chris Caldwell, and Marie Schaefer. "Indigenous Lessons about Sustainability Are Not Just for 'All Humanity.'" In *Sustainability*, edited by Julie Sze, 149–79. New York: New York University Press, 2018.

Wieviorka, Annette. *The Era of the Witness*. Translated by Jared Stark. Ithaca, NY: Cornell University Press, 2006.

Wildcat, Daniel R. *Red Alert! Saving the Planet with Indigenous Knowledge*. Golden, CO: Fulcrum Publishing, 2009.

Wiles, Siouxsie. "The Three Phases of Covid-19—and How We Can Make Them Manageable." *The Spinoff*, March 9, 2020. https://thespinoff.co.nz/society/09-03-2020 /the-three-phases-of-covid-19-and-how-we-can-make-it-manageable/.

Wilson, Kirt H. "Theory/Criticism: A Functionalist Approach to the 'Specific Intellectual' Work of Rhetorical Criticism." *Western Journal of Communication* 84, no. 3 (May 26, 2020): 280–96.

Wilson, Mark. "The Story Behind 'Flatten the Curve,' the Defining Chart of the Coronavirus." *Fast Company*, March 13, 2020. https://www.fastcompany.com/90476143 /the-story-behind-flatten-the-curve-the-defining-chart-of-the-coronavirus.

Winderman, Emily, Robert Mejia, and Brandon Rogers. "'All Smell Is Disease': Miasma, Sensory Rhetoric, and the Sanitary-Bacteriologic of Visceral Public Health." *Rhetoric of Health and Medicine* 2, no. 2 (Spring 2019): 115–46.

Wirtz, Kristina. "The Living, the Dead, and the Immanent: Dialogue across Chronotopes." *HAU: Journal of Ethnographic Theory* 6, no. 1 (June 1, 2016): 343–69.

Wirtz, Kristina. *Performing Afro-Cuba: Images, Voice, Spectacle in the Making of Race and History*. Chicago: University of Chicago Press, 2014.

Witze, Alexandra. "The Arctic Is Burning Like Never Before—and That's Bad News for Climate Change." *Nature* 585 (September 10, 2020): 336–37.

Wohlleben, Peter. *The Heartbeat of Trees: Embracing Our Ancient Bond with Forests and Nature*. Vancouver, BC: Greystone Books, 2021.

Wohlleben, Peter. *The Hidden Life of Trees: What They Feel, How They Communicate: Discoveries from a Secret World*. The Mysteries of Nature Series. Vancouver, BC, and Berkeley, CA: Greystone Books and the David Suzuki Institute, 2016.

Wong, Brittany. "13 Words or Phrases that Would Totally Confuse Pre-COVID Us." *Huffington Post*, December 18, 2020.

Woods, Derek. "Scale Critique for the Anthropocene." *Minnesota Review* 83 (2014): 133–42.

Woods, Derek. "Scale in Ecological Science Writing." In *The Routledge Handbook of Ecocriticism and Environmental Communication*, edited by Scott Slovic, Swarnalatha Rangarajan, and Vidya Sarveswaran, 118–28. New York: Routledge, 2019.

World Health Organization, Regional Office for South-East Asia. "Tuberculosis: 'Bending the Curve.'" Institutional Repository for Information Sharing, July 18, 2017. https://apps.who.int/iris/handle/10665/258547.

"Word of the Year 2019." *Oxford Languages*, n.d. Oxford: Oxford University Press, 2022. https://languages.oup.com/word-of-the-year/2019/.

Wright, C. D. *Casting Deep Shade: An Amble Inscribed to Beech Trees & Co.* Port Townsend, WA: Copper Canyon Press, 2019.

Wynn, James, and G. Mitchell Reyes. "From Division to Multiplication: Uncovering the Relationship Between Mathematics and Rhetoric through Transdisciplinary Scholarship." In *Arguing with Numbers: The Intersections of Rhetoric and Mathematics*, edited by James Wynn and G. Mitchell Reyes. University Park: Pennsylvania State University Press, 2021.

Wynn, James, and G. Mitchell Reyes. "Introduction." In *Arguing with Numbers: The Intersections of Rhetoric and Mathematics*, edited by James Wynn and G. Mitchell Reyes. University Park: Pennsylvania State University Press, 2021.

Wysocki, Anne Frances. "Unfitting Beauties of Transducing Bodies." In *Rhetorics and Technologies: New Directions in Writing and Communication*, edited by Stuart A. Selber. Studies in Rhetoric & Communication. Columbia: University of South Carolina Press, 2012.

Yazzie, Jaime O., Peter Z. Fulé, Yeon-Su Kim, and Andrew Sánchez Meador. "Diné Kinship as a Framework for Conserving Native Tree Species in Climate Change." *Ecological Applications* 29, no. 6 (September 1, 2019): 1331–42.

Ye, Zi-Wie, Shuofeng Yuan, Kit-San Yuen, Sin-Yee Fung, Chi-Ping Chan, and Dong-Yan Jin. "Zoonotic Origins of Human Coronaviruses." *International Journal of Biological Sciences* 16, no. 10 (2020): 1686–97.

Zalasiewicz, Jan, Colin Waters, and Martin J. Head. "Anthropocene: Its Stratigraphic Basis." *Nature* 541, no. 7637 (January 1, 2017): 289.

Zaval, Lisa, and James F. M. Cornwell. "Cognitive Biases, Nonrational Judgments, and Public Perceptions of Climate Change." In *Oxford Research Encyclopedia*. Oxford: Oxford University Press, 2017.

Zekollari, H., M. Huss, and D. Farinotti. "Modelling the Future Evolution of Glaciers in the European Alps under the EURO-CORDEX RCM Ensemble." *The Cryosphere* 13, no. 4 (2019): 1125–46.

Zelizer, Barbie. *About to Die: How News Images Move the Public.* Oxford: Oxford University Press, 2010.

Zelizer, Barbie. *Remembering to Forget: Holocaust Memory Through the Camera's Eye.* Chicago: University of Chicago Press, 1998.

Zemp, Michael, Holger Frey, Isabelle Gärtner-Roer, Samuel U. Nussbaumer, Martin Hoelzle, Frank Paul, Wilfried Haeberli, et al. "Historically Unprecedented Global Glacier Decline in the Early 21st Century." *Journal of Glaciology* 61, no. 228 (2015): 745–62.

Zuckerman, Andrew. "Andri Snær Magnason on How Time and Water Explain the Climate Crisis." *Time Sensitive* (podcast), episode 10, n.d. https://timesensitive.fm/episode/andri-snaer-magnason-time-water-climate-crisis/.

INDEX

Page numbers set in italics refer to illustrations or their captions.

Printed in Great Britain
by Amazon

24238386R00138